XXXVII Symposium on Nuclear Physics 2014

AA008590

Journal of Physics: Conference Series
Volume 578

Cocoyoc, Mexico
6 - 9 January 2014

Editors:

Roelof Bijker
Sergio Lerma
David Lizcano

ISBN: 978-1-63439-911-1
ISSN: 1742-6588

Printed from e-media with permission by:

Curran Associates, Inc.
57 Morehouse Lane
Red Hook, NY 12571

Some format issues inherent in the e-media version may also appear in this print version.

Copyright© by the Institute of Physics
All rights reserved.

Printed by Curran Associates, Inc. (2015)

For permission requests, please contact the Institute of Physics
at the address below.

Institute of Physics
Dirac House, Temple Back
Bristol BS1 6BE UK

Phone: 44 1 17 929 7481
Fax: 44 1 17 920 0979

techtracking@iop.org

Additional copies of this publication are available from:

Curran Associates, Inc.
57 Morehouse Lane
Red Hook, NY 12571 USA
Phone: 845-758-0400
Fax: 845-758-2634
Email: curran@proceedings.com
Web: www.proceedings.com

Table of contents

Volume 578

XXXVII Symposium on Nuclear Physics
6–9 January 2014, Cocoyoc, Mexico

Accepted papers received: 5 December 2014
Published online: 22 January 2015

Preface

Papers

 doi:10.1088/1742-6596/578/1/011001

Preface

The Symposium on Nuclear Physics is an international meeting organized every year since 1978 by the Division of Nuclear Physics of the Mexican Physical Society. The 37th edition was held at the Hotel Hacienda Cocoyoc in the state of Morelos, Mexico, from 6-9 January, 2014. The symposium is intended to be a relatively small meeting designed to bring together some of the leading nuclear scientists in different areas of nuclear physics (nuclear astrophysics, nuclear structure, neutrino physics, hadron physics and nuclear reactions among others). Both theorists and experimentalists, students, postdocs and senior scientists gathered in a relaxed and informal environment providing them with a unique opportunity to exchange ideas. The symposium was organized in plenary sessions with invited talks and a poster session.

In this edition the program consisted of 26 invited talks and 19 posters, whose subjects reflected the active areas and interests of the mexican nuclear physics community. The organization encouraged the participation of young scientist as speakers in the plenary sessions. Seven of the 26 invited talks were given by postdoctoral and doctoral students working in different institutions from Mexico, USA and Europe. The proceedings collect a total of 16 manuscripts from the invited speakers.

We acknowledge the enthusiastic collaboration of the **International Advisory Committee** formed by

- Osvaldo Civitarese (Universidad Nacional de La Plata, Argentina)
- Jerry Draayer (Louisiana State University, USA)
- Alfredo Galindo-Uribarri (Oak Ridge National Laboratory, USA)
- Paulo Gomes (Universidade Federal Fluminense, Brazil)
- Piet Van Isacker (GANIL, France)
- Reiner Krücken (TRIUMF, Canada)
- Jorge López (The University of Texas at El Paso, USA)
- Jorge Piekarewicz (Florida State University, USA)
- Stuart Pittel (University of Delaware, USA)
- W. Michael Snow (Indiana University, USA)
- Adam Szczepaniak (Indiana University, USA)
- Michael Wiescher (University of Notre Dame, USA)

The **local organizing committee** was formed by

- Roelof Bijker (Instituto de Ciencias Nucleares, UNAM)
- Sergio Lerma (Facultad de Física e Inteligencia Artificial, UV)
- David Lizcano (Instituto Nacional de Investigaciones Nucleares, SENER)
- Irving Morales (Instituto de Ciencias Nucleares, UNAM)

- **Invited talks**
  - **Ani Aprahamian** (University of Notre Dame, USA) *Sensitivity of the r-process to nuclear masses and beta decay rates in three astrophysical scenarios.*
  - **Libertad Barrón** (for the NPDGamma Collaboration, Instituto de Física-UNAM, Mexico) *Low energy-neutrons in the study of the Hadronic Weak Interaction: The NPDGamma Experiment*
  - **Roelof Bijker** (Instituto de Ciencias Nucleares-UNAM, Mexico) *Valence and sea quarks in the nucleon*
  - **Efraín Chávez** (Instituto de Física-UNAM, Mexico) *Nuclear (astro)physics in the upgraded accelerator facilities at IFUNAM.*
  - **Osvaldo Civitarese** (Universidad Nacional de la Plata, Argentina) *On the definition of entropy for unstable states*
  - **Eduardo Coello Pérez** (University of Tennessee, USA) *Electric transitions of deformed nuclei within an effective theory*
  - **Jerry Draayer** (Louisiana State University, USA) *Dominant Modes in Light Nuclei. Ab Initio View of Emergent Symmetries*
  - **Francisco Favela** (Instituto de Física-UNAM, Mexico) *Progress in the characterization of the gas jet target.*
  - **Ruben Fossion** (Instituto Nacional de Geriatría and Centro de Ciencias de la Complejidad-UNAM, Mexico) *Data-adaptive unfolding of nonergodic spectra in spectral fluctuation studies*

Journal of Physics: Conference Series **578** (2015) 011001 doi:10.1088/1742-6596/578/1/011001

- **Arturo Gómez-Camacho** (Instituto Nacional de Investigaciones Nucleares, Mexico)*CDCC calculations of elastic scattering angular distributions for 6 Li with several targets using global potentials*
- **Peter Hess** (Instituto de Ciencias Nucleares-UNAM, Mexico) *Neutron stars with dark energy*
- **Tonatiuh Jiménez-Bustamante** (Instituto de Ciencias Nucleares-UNAM, Mexico) *Identified particle production in pp, p-Pb, and Pb-Pb collisions measured with ALICE at the LHC energies*
- **Arturo Menchaca** (Instituto de Física-UNAM, Mexico) *Asymmetric vs. symmetric heavy ion fusion: an old riddle solved using macroscopic liquid drop collisions.*
- **Lizeth Morales-Gallegos** (University of Edinburgh, Scotland) *12C+12C reactions at astrophysical energies: Tests of targets behaviour under beam bombardment*
- **Elizabeth Padilla** (Instituto de Ciencias Nucleares-UNAM Mexico) *Spectroscopy of neutron-rich nuclei*
- **Edgar Pérez-Lezama** (Instituto de Ciencias Nucleares-UNAM, Mexico) *Study of the Pythia 8 delta eta delta phi correlation in function of multiplicity and transverse sphericity in proton proton collisions at 7 TeV*
- **Stuart Pittel** (University of Delaware, USA) *Exact solution of the pairing hamiltonian in atomic nuclei*
- **W. Michael Snow** (Indiana University, USA) *Searching for Dark Energy Using Neutron Interferometry.*
- **Scott Suchyta** (NSCL, Michigan State University USA) *Shape coexistence in 68Ni*
- **Adam Szczepaniak** (Indiana University, USA) *Hadron Spectroscopy: past, present and future.*
- **Carlos Vargas** (Universidad Veracruzana, Mexico) *Microscopic study of collectivity in 160170Dy nuclei*
- **Victor Velázquez** (Facultad de Ciencias-UNAM, Mexico) *Complexity, decoherence and quantum chaos in the nuclear shell model.*
- **Michael Wiescher** (University of Notre Dame, USA) *Nuclear Astrophysics Challenges at Threshold Energies.*
- **Kathrin Wimmer** (Central Michigan University, USA) *Single-particle structure of neutron-rich N=40 isotopes.*
- **Tochtli Yépez** (Indiana University, USA) *Charmonium meson and hybrid radiative transitions*
- **Kai Zuber** (Technische Universitt Dresden, Germany) *Status and perspectives of double beta decay*

- Posters
  - **Arturo Amor** (Instituto de Ciencias Nucleares-UNAM, Mexico) *A Nonperturbative-QCD oriented variational procedure*
  - **Victoria Isabel Araujo Escalona** (Instituto de Física-UNAM, Mexico) *The nucleosynthesis of LEMA, the case of the 26Al.*
  - **María Guadalupe Caldera Cabral** (Universidad Autónoma de Zacatecas, Mexico) *Determination of the parameters alpha and f with the implementation of* $K_0 - NAA$ *method, with two monitors in the Triga Mark III Reactor*
  - **Silvia Diaz Gómez** (Instituto de Ciencias Nucleares-UNAM, Mexico) *Baryon magnetic moments in three models of quarks effective*
  - **Hugo García Tecocoatzi** (Instituto de Ciencias Nucleares-UNAM, Mexico) *Electromagnetic decays in Baryons*
  - **Araceli Ibañez Sandoval** (ESIME-IPN, Mexico) *Study of Yrast structure in the even-mass 128-132 Nd nuclei with the Projected Shell Model*

Journal of Physics: Conference Series **578** (2015) 011001 doi:10.1088/1742-6596/578/1/011001

- **Daniel Julián Nader** (Instituto de Ciencias Nucleares-UNAM, Mexico) *Mean field in the Pairing plus Quadrupole Model: BCS aproximation compared with Richardson solution in s-d and p-f shells*
- **José Lozano Torres** (Universidad Veracruzana, Mexico) *Isovectorial pairing plus quadrupole model in the framework of SU(3) scheme*
- **Ruslan Magana-Vsevolodovna** (Universidad Juarez Autónoma de Tabasco, Mexico) *Intensities of two nucleon transfer reaction between 194Au and 194Pt*
- **Marissa Maldonado Velázquez** (Instituto de Física-UNAM, Mexico) *Magnetic fields for the transport and manipulation of low energy neutrons's spin*
- **Enrique Martínez Quiroz** (Instituto Nacional de Investigaciones Nucleares, Mexico) *Complete and incomplete fusion of the* ^{6}Li+^{58}Ni *system*
- **Giovani Erick Morales Hernández** (Instituto de Ciencias Nucleares-UNAM, Mexico) *Phase transitions for excited states in light nuclei within the Semimicroscopic Algebraic Cluster Model (SACM) using the formalism of catastrophe theory.*
- **Juan Carlos Morales Rivera**(UAEMex-Instituto Nacional de Investigaciones Nucleares, Mexico) *Optical model analysis for the* ^{9}Be+ ^{27}Al *system*
- **Emmanuel Ortiz** (Instituto de Ciencias Nucleares-UNAM, Mexico) *Baryon beta decays: sea and valence quarks*
- **Adrián Pallares** (Instituto de Física-UASLP, Mexico) *Geometric so(4) symmetry loss without degeneracy lift*
- **Victor Olenin Ramírez Beltrán** (Universidad Autónoma del Estado de México, Mexico) *Fusion-evaporation analysis for* ^{7}Li *and* ^{7}Be *mirror nuclei*
- **Carlos Alberto Ramírez Medina** (Universidad Autónoma del Estado de México, Mexico) *Dynamics of the nonlinear wave packets in the harmonic oscillator potential*
- **Andrés Ramírez Morales** (Instituto de Física-UNAM, Mexico) *Magnetic fields for neutron spin manipulation and transport in the n-3He experiment*
- **Francisco Ramírez Sánchez** (Universidad Autónoma de Zacatecas UAEN-UAP, Mexico) *Natural Gamma radioactivity level analysis in rubbish (mining waste) from mineral mining of the Fresnillo PLC company in Mexico*

Roelof Bijker
Sergio Lerma
David Lizcano
Editors

Journal of Physics: Conference Series **578** (2015) 011002 doi:10.1088/1742-6596/578/1/011002

Peer review statement

All papers published in this volume of *Journal of Physics: Conference Series* have been peer reviewed through processes administered by the proceedings Editors. Reviews were conducted by expert referees to the professional and scientific standards expected of a proceedings journal published by IOP Publishing.

Progress in the characterization of the gas jet target at IFUNAM

F Favela, E Chávez, O de Lucio, E Andrade and M E Ortíz

Instituto de Física, Universidad Nacional Autónoma de México Apartado postal 20-364, 01000, México, D.F., México

E-mail: ffavela@gmail.com

Abstract.
The windowless gas target (supersonic jet) at IFUNAM has been characterized using the Elastic Backscattering Spectrometry (EBS) technique with proton beams. Jets of Air, Nitrogen and Argon at different pressures were studied. Areal densities above 10^{18} atoms/cm^2 have been achieved.

1. Introduction

An ideal target for experiments in nuclear physics would be one that has a large number of the target nucleus of interest and no others (dense), in a well-defined region of space (thin) and finally that it does not change its properties with beam exposure over time.

A supersonic gas jet closely resembles such an ideal target. Systems like this have been used in the past [1, 2, 3] and a more contemporary example has been developed in the University of Notre Dame [4].

The 5.5 MV Van de Graaff accelerator laboratory at IFUNAM (Instituto de Física Universidad Nacional Autónoma de México) underwent in 2012-3 a significant upgrade in the experimental area. The addition of a switching magnet allowed the number of beam lines to increase from 1 to 7. One of these has been designated to host the supersonic gas jet target. The scientific program behind its design and construction is mainly related with nuclear physics and astrophysics research. Although Atomic physics and Material Sciences studies are also possible.

2. Experimental methods

Figure 1 shows a picture of the setup for the jet characterization experiments. Inside the scattering chamber a special stainless steel table was made to support charged particle detectors. The table is held from the central catcher and it has holes in order to preserve the pumping speed of the system. On this table, two Silicon charged particle detectors are placed, left and right, $15cm$ away from the gas target at an angle of 150 degrees relative to the beam direction.

Three millimeter in aperture diameter and one centimeter long, collimators where placed in front of the detectors in order to shield the detector from the scattering of the beam as it goes through the scattering chamber before and after the jet (the surrounding region is not perfect vacuum). The solid angle for each detector, with their collimator, is $\Omega \approx 4 \times 10^{-4} sr$ and the pressure in the surrounding region of the jet was $\approx 10^{-1}$ Torr.

Journal of Physics: Conference Series **578** (2015) 012001 doi:10.1088/1742-6596/578/1/012001

Figure 1. EBS setup. During an experiment the beam should go from left to right.

3. Results

For the first test, we chose the easiest target; air. The nozzle's valve was simply opened to let in air from the laboratory (a huge leak under control). A $2.8MeV$ proton beam ($\approx 1\mu A$) was used to produce EBS data to characterize our target.

Figure 2 shows two of the spectra taken. Both show the two expected peaks from proton scattering on the principal components of air: N and O. The spectrum on the left however has a significant count yield at lower and higher energies. The spectrum on the right hand side was taken in the same conditions using the collimator described above and shows a significant reduction of this background.

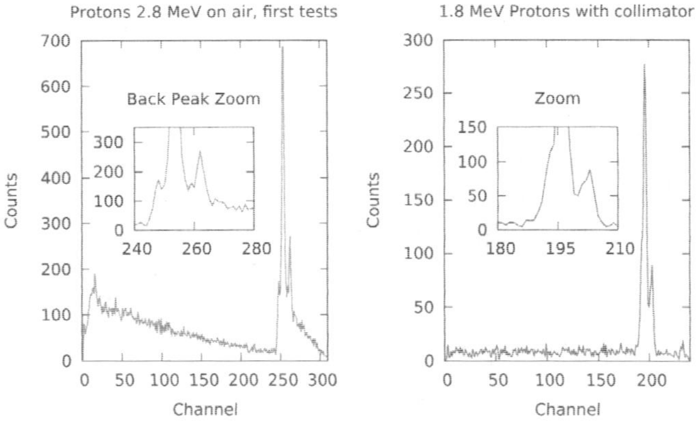

Figure 2. Experimental comparison between the data without the collimator (left) and with the collimator (right)

EBS data was analyzed using the SINMRA software [5]. This program takes advantage of a rather large cross section data base (IBANDL [6]) to simulate EBS spectra for a given experimental setup (beam, detector and target configuration). It is through this analysis that we can extract both target composition and density. Table 1 shows the elemental abundance of the main components of air extracted from our data in comparison with typical reported values [7].

Using the same experimental protocol, we proceed to characterize gas jet targets of pure gases: Ar and N. This time, SIMNRA analysis is not needed to extract the elemental composition, but to yield the target density, a parameter of utmost importance for our target.

Figure 3 shows the evolution of the areal density with the input pressure of the gas bottles. There is a nearly proportional dependence similar in both gases. However, the main result is the absolute value of the areal density obtained which is in excess of 10^{18} particles$/cm^2$.

Peak	SIMNRA	reported
Nitrogen	78.0%	78.09%
Oxygen	21.3%	20.95%
Argon	0.97%	0.93%

Table 1. Comparison of the SIMNRA simulation and the reported percentages in Air.

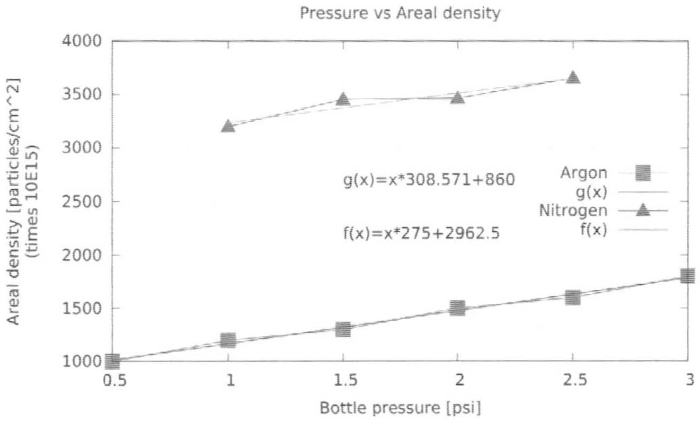

Figure 3. Areal density evolution with pressure in nitrogen and argon.

4. Conclusions

In this work we showed the performance of the supersonic gas jet target recently built at IFUNAM.

Targets of Air, Nitrogen and Argon were studied and characterized using EBS data with proton beams and data reduction with the SIMNRA code.

The main result is that we showed that we can reach jet densities in excess of 10^{18} particles/cm^2.

The more intense ion source planed in the near future for the $5.5MV$ accelerator will enhance the capability of the supersonic gas jet target beam line to its full potential and take on astrophysical experiments.

5. Acknowledgments

Special thanks to Marco Veytia, Roberto Gleason, Hesiquio Várgas, and IFUMAN's shop.

This work was done thanks to the support of CONACYT 51600, 82692, 123655 and DGAPAUNAM IN-118310, IN-112609, IN-105510, IN-115213, TA-100213, IN-103312.

References

[1] Shapira D, Jr J F, Novotny R, Shivakumar B, Parks R and Thornton S 1985 *Nucl. Instr. and Meth. in Physics Research Section A: Accelerators, Spectrometers, Detectors and Associated Equipment* **228** 259 – 266 ISSN 0168-9002 URL `http://www.sciencedirect.com/science/article/pii/0168900285902670`

[2] Shapira D, Campo J D, Jr J F, Shivakumar B, Stelson P, Harmon B, Parks R and Thornton S 1985 *Nucl. Instr. and Meth. in Physics Research Section B: Beam Interactions with Materials and Atoms* **1011, Part 1** 436 – 440 ISSN 0168-583X URL `http://www.sciencedirect.com/science/article/pii/0168583X8590285X`

[3] Becker H, Buchmann L, Gorres J, Kettner K, Krawinkel H, Rolfs C, Schmalbrock P, Trautvetter H and Vlieks A 1982 *Nucl. Instr. and Meth. in Phys. Res.* **198** 277–292 ISSN 0167-5087 URL `http://www.sciencedirect.com/science/article/pii/0167508782902654`

[4] Kontos A, Schurmann D, Akers C, Couder M, Gorres J, Robertson D, Stech E, Talwar R and Wiescher M 2012 *Nucl. Instr. and Meth. in Phys. Res. S. A* **664** 272 – 281 ISSN 0168-9002 URL http://www.sciencedirect.com/science/article/pii/S0168900211019899

[5] Mayer M 1999 *American Institute of Physics Conference Series* (*American Institute of Physics Conference Series* vol 475) pp 541–544

[6] Gurbich A 2014 Ion beam nuclear data library (IBANDL) web application for cross sections URL https://www-nds.iaea.org/exfor/ibandl.htm

[7] NASA 2013 Earth fact sheet typical air abundance variables URL http://nssdc.gsfc.nasa.gov/planetary/factsheet/earthfact.html

Journal of Physics: Conference Series **578** (2015) 012002 doi:10.1088/1742-6596/578/1/012002

Tests of carbon targets for ^{12}C+^{12}C reactions at astrophysical energies

L Morales-Gallegos[1,2], M Aliotta[1], A Di Leva[2,3], L Gialanella[2,4], D Schürmann[2], M De Cesare[2,5], T Davinson[1], G Imbriani[2,3], M Romano[2,3] and M Romoli[2,3]

[1] SUPA, School of Physics and Astronomy, University of Edinburgh, Edinburgh, UK
[2] INFN, Sezione di Napoli, Naples, Italy
[3] Dipartimento di Scienze Fisiche, Università di Napoli "Federico II", Naples, Italy
[4] Dipartimento di Matematica e Fisica Seconda Università degli studi di Napoli, Caserta, Italy
[5] Department of Nuclear Physics, Research School of Physics and Engineering, Australian National University, Canberra, ACT 0200, Australia

E-mail: `Lizzy.Morales@ed.ac.uk`

Abstract. As a preliminary step towards measurements of the ^{12}C +^{12} C reactions at astrophysical energies, we investigate the behaviour of targets under beam bombardment, specifically the quantitative relation between hydrogen and deuterium content of different carbon targets and target temperature. Experiments have taken place at the CIRCE accelerator in Caserta, Italy and preliminary results are presented here.

1. Introduction

^{12}C + ^{12}C reactions in stars take place in the carbon burning stage of stellar evolution at a typical temperature of 5x10^8 K that corresponds to an energy of $E_0 = 1.5 \pm 0.3$ MeV [1, 2, 3]. At these energies, the carbon burning proceeds through the reactions ^{12}C(^{12}C, α)^{20}Ne and ^{12}C(^{12}C, p)^{23}Na, known as the α and p channels, respectively.

The way in which carbon is burned depends critically on the mass of the star and it is determined by the rate of the ^{12}C+^{12}C reactions. For this reason, the rate of ^{12}C+^{12}C reactions is one of the key quantities needed to understand the evolution of massive stars (>8M$_\odot$) and the nucleosynthesis of heavy elements [4, 5].

Early investigations of the ^{12}C + ^{12}C reactions at low energies go back to 1969. To measure these reactions, different groups have used charged particle detection [6, 7, 8] and gamma spectroscopy techniques [9, 10, 11, 12, 13, 14, 15, 16] to investigate both the α and p channels. To date, the measurement that reached the lowest energy has been the one of Spillane *et al.* with an energy of $E = 2.10$ MeV [16]. The current situation of ^{12}C +^{12} C reactions measurements is presented in figure 1 extracted from [17]. In this figure, the problems at measuring the ^{12}C+^{12}C reactions are pictured: there are not well understood discrepancies between the different data sets, measurements are still far from the stellar energy range, extrapolations to these energies are very uncertain since the resonant structure of these reactions lead to different theoretical models that predict astrophysical S-factors orders of magnitude different between each other and measurements below $E = 3$ MeV have significant error bars due to contaminants in the

Journal of Physics: Conference Series **578** (2015) 012002 doi:10.1088/1742-6596/578/1/012002

targets [6, 7, 14, 16]. These contaminants are mainly hydrogen and deuterium. When the ^{12}C beam impinges in the ^{12}C target, the probability of producing a $^{12}C+^{1}H$ reaction is orders of magnitude greater than the probability of having a $^{12}C+^{12}C$ reaction because the Coulomb barrier of $^{12}C+^{1}H$ is significantly lower than the one for $^{12}C+^{12}C$. Deuterium contributes to the unwanted background by the so called "two step process" [18]. In this process, deuterium is elastically scattered by the ^{12}C beam at forward angles and then impinges on a ^{12}C in the target producing a background proton.

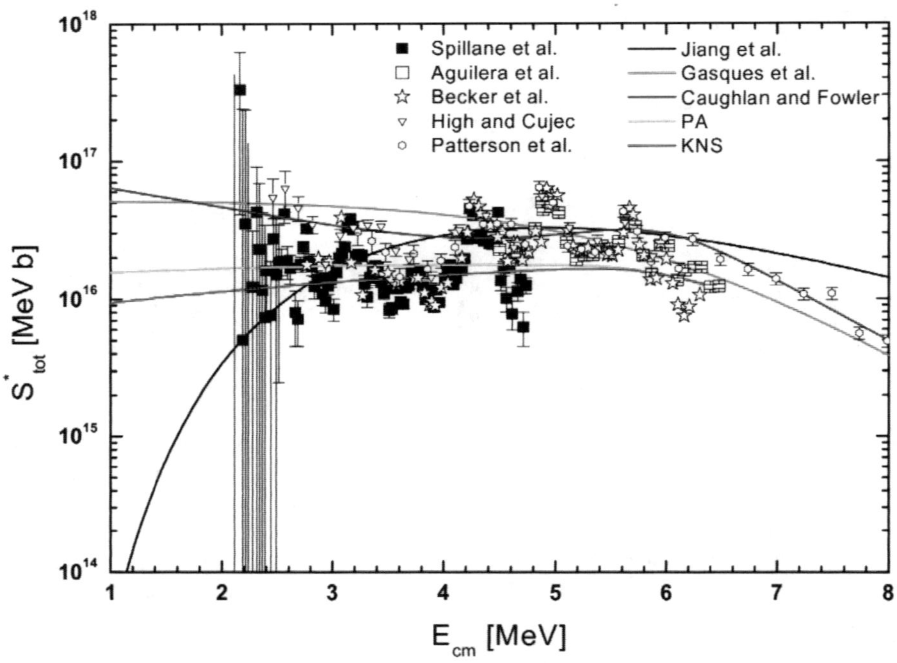

Figure 1. Present situation of the $^{12}C+^{12}C$ reactions measurements. Astrophysical S*-factor for the $^{12}C + ^{12}C$ reactions as a function of centre of mass energy. Open and filled symbols represent experimental data points (errors are statistical only); curves represent different theoretical models.

Some groups [14, 16, 10] have tried to reduce this target contamination by heating the targets up in different ways, either using resistance heating to increase the temperature to 1800° C of carbon foils ($9-88$ $\mu g/cm^2$) placed on a tantalum backing [10] or beam heating thick (1 mm) graphite targets at 600° C (for 6-8 hours) [14] and 700° C [16]. In all cases, a reduction in hydrogen content of the targets was found, resulting in cleaner spectra at low energies. Nevertheless, this reduction was never quantified because of time or equipment limitations. Thus, the aim of this work is to quantify the time variation of hydrogen and deuterium content in a target as a function of target temperature using a systematic approach.

2. Experimental set up and method

The experimental set up is schematically shown in figure 2. Briefly, it consists in a small chamber housing a water cooled target holder, a detector holder, a detector and a cold finger with an Al foil in front of the detector. The target holder can accommodate two different targets and two collimators of 3 and 6 mm to focus the beam, it is water cooled in order to be able to control the temperature of the target and minimize its degradation. A ($\Delta E-E_{rest}$) telescope Si

Journal of Physics: Conference Series **578** (2015) 012002 doi:10.1088/1742-6596/578/1/012002

detector (25μm and 300μm, respectively) is placed at 135° with respect to the beam axis; this type of detector was chosen in order to allow for particle identification. A cold finger to control carbon build-up was used and kept at a potential of -300V to suppress secondary electrons. The cold finger holds a 2μm Al foil placed in front of the detector to protect it from elastically scattered particles. A quadrupole mass spectrometer (QMS) was used to continuously monitor the composition of the residual gas since the atoms it contains can be easily deposited on the surface of the target increasing its contamination. A FLIR SC325 calibrated thermocamera (accuracy of ±2% at reading) [19] was placed in a chamber's viewport looking through a Ge window. This window is transparent to the wavelengths the thermocamera is sensitive to (7.5μm-13μm).

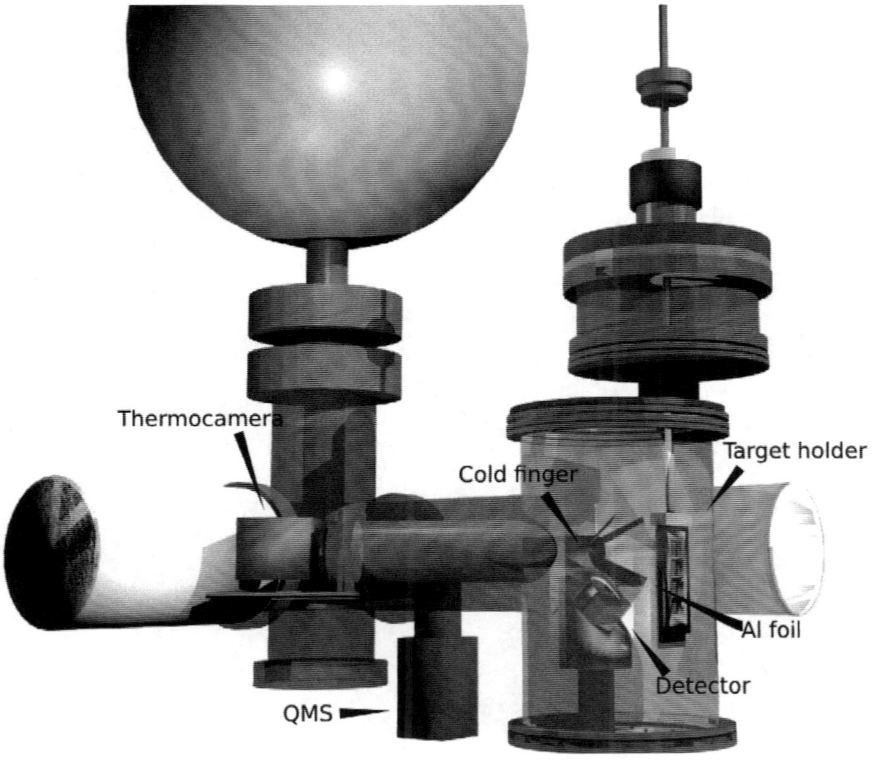

Figure 2. Sketch of the experimental set up. Shown are the detector, the target holder, the cold finger, the thermocamera placed outside the chamber in front of a Ge window and the quadrupole mass spectrometer.

Since the purpose of these tests is to quantitatively determine the ^{1}H and ^{2}H content of our targets as a function of temperature during ion beam bombardment, we employ Nuclear Reaction Analysis (NRA) to quantify the contaminants in the target. We use C, N and O beams coming from the 3 MV pelletron tandem accelerator at CIRCE (Centre for Isotopic Research on the Cultural and Evironmental heritage) laboratory in Caserta, Italy. The thermocamera outside the chamber is used to monitor the target temperature continuously. The thermocamera measurements are attenuated by the Ge window but this attenuation is well known (calibration certificate No. SED04024 by FLIR) and taken into account during the analysis.

For visual purposes, an image taken with the thermocamera facing the interior of the chamber is shown in figure 3. We can clearly identify a part of the target holder and the cold finger as well as the beam on the target in this figure.

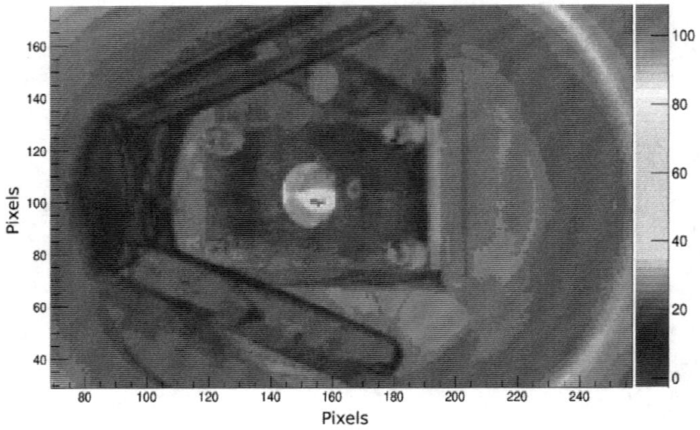

Figure 3. Thermocamera picture of target ladder with a beam on the target. The colour bar on the right shows the temperature scale in °C.

3. Preliminary results

Initial tests were made with a $^{12}C^{+3}$ ion beam of $E = 4$ MeV and an intensity of 7 particle μA. The beam hit an infinitely thick target (1mm). From the images acquired with the thermocamera it was possible to derive the size of the beam and found it was less than 2.5 mm. The beam constantly heated the target at the beam spot, reaching a temperature of about 500° C in 50 minutes (figure 4).

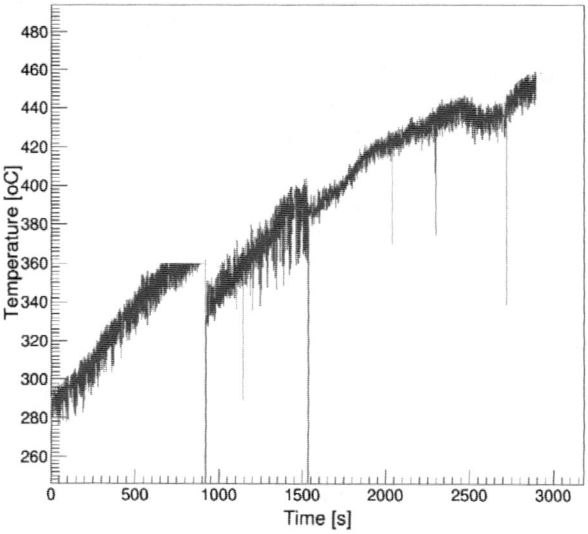

Figure 4. Evolution of target temperature (at the beam spot position) as a function of time. Discontinuities in the trend correspond to changes in the temperature range settings of the thermocamera.

A preliminary $\Delta E - E$ matrix is shown in figure 5. The axes correspond to the ΔE detector energy and the total energy ($\Delta E + E_{rest}$). The most prominent groups in the locus shown correspond to the proton groups from the $^{12}C(^{12}C, p)^{23}Na$ reaction, with p_0 corresponding to reactions with ^{23}Na left in its ground state and p_1 to reactions with ^{23}Na left in its first excited state. The total energies of p_0 and p_1 proton groups are in agreement with the kinematically calculated values. At energies lower than 3 MeV, background events from $^{12}C(d, p)$ reactions due to contaminants in the target were observed as expected given that we did not use highly pure targets for this test. High energy protons can also be seen and presumably correspond to reactions with ^{13}C contaminants in the target, although this is still under study.

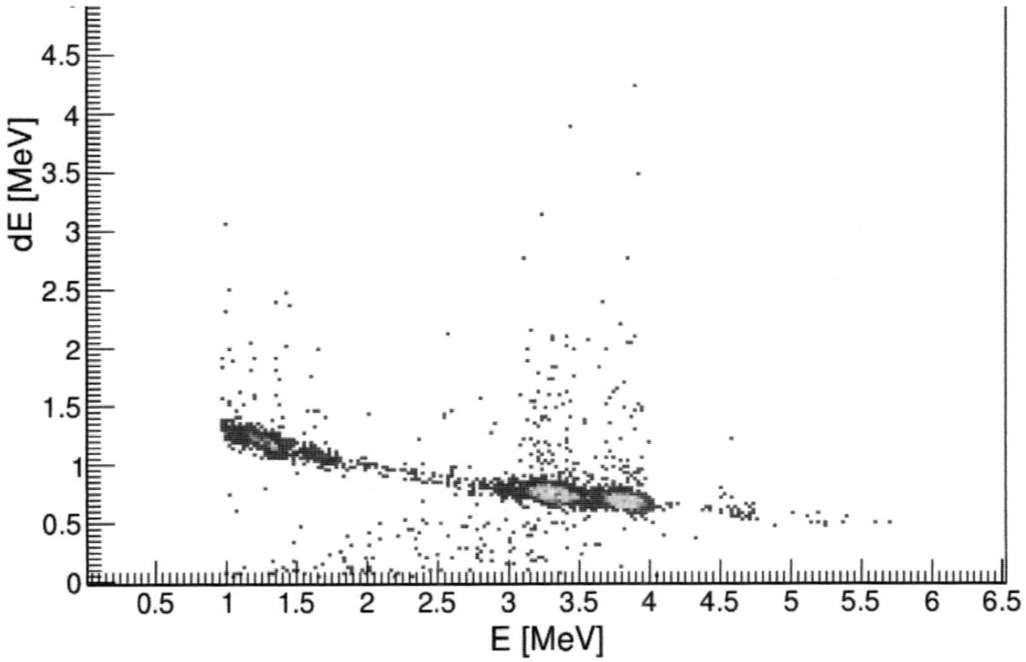

Figure 5. $\Delta E - E$ matrix from the telescope detector. The axis correspond to the ΔE detector energy and the total energy ($\Delta E + E_{rest}$). The most intense groups between 3 and 4 MeV correspond to the $^{12}C(^{12}C, p)^{23}Na$ reaction; protons with lower energies come from deuterium contamination in the target and high energetic protons possibly correspond to ^{13}C contaminants in the target.

In order to quantify the H content of the targets we will use targets with known H concentration. The hydrogen content of the target will be continuously monitored during the tests of the target behaviour under beam bombardment. Measurements will be repeated with different beam intensities, thus allowing us to determine the time evolution of the target's H content as a function of target temperature.

4. Summary and outlook

We have shown that the designed setup is able to provide a temperature map of the beam impact area on the C target and, concurrently, a measurement of the target's H content. This approach could be extended to other light isotopes, possible sources of beam induced background. Comparative tests of different target materials will be performed to determine the best conditions for the measurements of the $^{12}C + ^{12}C$ fusion reactions. We will also study the composition of

Journal of Physics: Conference Series **578** (2015) 012002 doi:10.1088/1742-6596/578/1/012002

the residual gas with the quadrupole mass spectrometer and put the chamber into a N aquarium to reduce light ions contamination.

References

[1] Strieder F and Rolfs C 2007 *Progress in Particle and Nuclear Physics* **59** 562–578
[2] Woosley S, Wunsch S and Kuhlen M 2004 *Astropysical Journal* **607** 921
[3] Baraffe I, Heger A and Woosley S 2004 *Astropysical Journal* **615** 378
[4] Gasques L, Afanasjev a, Aguilera E, Beard M, Chamon L, Ring P, Wiescher M and Yakovlev D 2005 *Physical Review C* **72** 025806
[5] Rolfs C and Rodney S *Cauldrons in the Cosmos* (Chicago and London: The University of Chicago Press)
[6] Patterson J, Winkler H and Zaidins C 1969 *Astrophysics journal* **157** 367
[7] Mazarakis M and Stephens W 1973 *Physical Review C* **7** 4
[8] Becker H W, Kettner K U, Rolfs C and Trautvetter H P 1981 *Zeitschrift für Physik A, Atoms and Nuclei* **312** 305–312
[9] High M and Ĉujec B 1977 *Nuclear Physics A* **282** 181–188
[10] Kettner K U 1980 *Zeitschrift für Physik A, Atoms and Nuclei* **75** 65–75
[11] Rosales P and Aguilera E 2003 *Revista Mexicana de Física S4* **49** 88–91
[12] Barrón-Palos L and Chavez E 2004 *Revista Mexicana de Física S* **50** 18–23
[13] Barrón-Palos L, Chávez L E, Huerta H a, Ortiz M E, Murillo O G, Aguilera R E, Martnez Q E, Moreno E, Policroniades R R and Varela G a 2005 *The European Physical Journal A* **25** 645–646
[14] Barrón-Palos L, Aguilera E, Aspiazu J, Huerta a, Martínez-Quiroz E, Monroy R, Moreno E, Murillo G, Ortiz M, Policroniades R, Varela a and Chávez E 2006 *Nuclear Physics A* **779** 318–332
[15] Aguilera E, Rosales P, Martinez-Quiroz E, Murillo G, Fernández M, Berdejo H, Lizcano D, Gómez-Camacho a, Policroniades R, Varela a, Moreno E, Chávez E, Ortíz M, Huerta a, Belyaeva T and Wiescher M 2006 *Physical Review C* **73** 064601
[16] Spillane T, Raiola F, Rolfs C, Schürmann D, Strieder F, Zeng S, Becker H W, Bordeanu C, Gialanella L, Romano M and Schweitzer J 2007 *Physical Review Letters* **98** 122501
[17] Strieder F 2008 *Journal of Physics G: Nuclear and Particle Physics* **35** 014009
[18] Zickefoose J, Schweitzer J, Spillane T, Strieder F, Becker H W, Rolfs C, Leva A D, Cesare M D, Cesare N D and Terrasi F 2010 *Proceedings of Science* pp 1–5
[19] http://wwwflircom/

Journal of Physics: Conference Series **578** (2015) 012003 doi:10.1088/1742-6596/578/1/012003

CDCC calculations of elastic scattering for the systems ^{6}Li+^{144}Sm and ^{6}Li+^{208}Pb. Effect of resonances of ^{6}Li on elastic scattering angular distributions

A Gómez Camacho[1]

[1]Departamento de Aceleradores, Instituto Nacional de Investigaciones Nucleares, Apartado Postal 18-1027, C.P. 11801, México, D.F., México

E-mail: `arturo.gomez@inin.gob.mx`

Abstract. Calculations of elastic scattering angular distributions for reactions of the weakly bound projectile ^{6}Li with targets ^{144}Sm and ^{208}Pb at energies above the barrier, are performed with the Continuum Discretized Coupled-Channel method (CDCC). Ground, resonant and non-resonant continuum states of ^{6}Li are included up to some maximum energy ϵ_{max} for which convergence is achieved. In the three-body system, global interactions are used for the $\alpha-target$ and $d-target$ sub-systems. The effect of continuum resonant states of ^{6}Li, *i.e.*, $l = 2$, $j^{\pi} = 3+, 2+$ and $1+$ on elastic scattering angular distributions is investigated by extracting these states from the continuum space. It is found that the calculated elastic scattering angular distributions are in good agreement with the measurements for most of the cases studied where consideration of couplings to continuum states is essential. It is also found that the resonance character of the continuum states is in some cases important to obtain agreement with the data.
PACS Numbers: 24.10Eq, 25.60Pj, 25.60Bx, 25.60Gc

1. Introduction

Reaction mechanisms involving weakly bound nuclei, both stable and radioactive, have been a subject of strong research[1, 2, 3]. One of the most interesting topics, is the effect of breakup on elastic scattering and fusion reaction mechanisms. It is now known that couplings of the elastic channel to continuum breakup states of the weakly bound projectile, as well as, continuum-continuum couplings are essential to fit the experimental data. This is so since, couplings to and between continuum states give rise to a strong repulsive polarization potential that decreases the reaction absorption. On the other side, besides the non-capture breakup process, other nuclear mechanisms are present for reactions with weakly bound nuclei. For instance, sequential and direct complete fusion, incomplete fusion and nucleon transfer. Coupled channel calculations show that couplings between the incident elastic channel to fusion ones, give place to an attractive dynamical polarization potential V_F that lowers the nominal Coulomb barrier V_B. So, the net effect of V_F is to enhance fusion mainly for energies around the barrier. Similarly, couplings to breakup channels produce a repulsive dynamical polarization potential V_{BU}, that is particularly strong at energies below the barrier. In this energy range, V_{BU} overcomes V_F, so that the net effect is twofold: a) An increase in the fusion barrier, that leads to a net fusion suppression, and b) Most of the reaction cross section is due to breakup. Several CDCC calculations have been proposed to calculate the effect of breakup couplings on elastic and fusion cross sections of several weakly bound systems. For instance, the $2n$-halo ^{6}He on ^{59}Co and ^{208}Pb[4, 5], ^{6}Li with targets ^{28}Si, ^{59}Co, ^{58}Ni, ^{144}Sm and ^{208}Pb[4, 6, 7, 8, 9, 10, 11, 12, 13], also ^{7}Li with ^{28}Si[14, 15] and ^{144}Sm[16], the n-halo ^{11}Be with ^{208}Pb[17] and the p-halo ^{8}B with ^{58}Ni[10, 18]. Most of the CDCC calculations in these works, use either, microscopic density dependent double-folding interactions with normalization factors N_R, N_I or Woods-Saxon potentials obtained from experimental systematic analyses. Except for a few cases, where contradictory results are found, e.g., ^{8}B+^{58}Ni[10, 18], most of the CDCC calculations show that couplings to continuum states of the weakly bound projectile have a strong effect and their consideration is very important to attain agreement with the experimental data. However, most of the studies for the weakly bound nucleus ^{6}Li do not distinguish the effect that the resonance character of the continuum states 3+, 2+ and 1+ can have on elastic scattering. The purpose of the present work is to calculate this effect on reactions with the spherical targets ^{144}Sm and ^{208}Pb, at energies above the Coulomb barrier. The CDCC[6, 19] method is used with global $d - target$ and $\alpha - target$ interactions, which depend only on the target mass A and incident collision energy E_{lab}. The global Woods-Saxon potentials, given in Ref.[20] for the $d - target$ interaction V_{dT}, are obtained from a large analysis of experimental elastic and inelastic cross sections for reactions of deuteron with a large variety of target masses ($12 < A < 238$) and range of incident energies $E_{lab} < 183$ MeV. As a matter of fact, this interaction is an improvement of other widely used systematic parametrizations for shorter range of masses and energies[21, 22, 23]. It is the purpose of the calculations performed in this work, to test this improved parametrization of the $d - target$ interaction, when used to describe continuum states of the deuteron after the breakup of ^{6}Li. The density dependent double-folding Sao Paulo potential (SPP)[24] with adequate α-particle mass densities is used to describe the $\alpha - target$ interaction $V_{\alpha T}$. The non-local SPP has been widely used with success to describe elastic and inelastic scattering, as well as, nucleon transfer cross sections for a number of nuclear systems. An extensive systematics of nuclear densities has been performed by L.C. Chamon *et al.*,[25] to produce a parameter-free interaction. In our calculations, the $\alpha - target$ interaction depends also only on the target mass A and the kinetic energy of the α-particle respect to the initial incident energy of ^{6}Li. As for the $\alpha - d$ cluster structure of ^{6}Li, use is made of the Woods-Saxon parametrization given in Ref.[26]. Ground, and unbound continuum scattering states of ^{6}Li are generated with this interaction. In order to calculate the effect of the resonances of ^{6}Li, we remove the continuum states $l = 2, s = 1, \pi = (-)^l$, that is j^π=3+, 2+ and 1+ from the continuum spectrum. This will be explained in section 3 where, the results are presented.

2. Basic CDCC formalism

In the CDCC formalism, continuum scattering states are replaced by a limited summation over bin-states. A bin-state N_l, $l = 1, ..., l_{max}$ is a set of discrete states, with a given angular momentum l, with parity $(-1)^l$, total angular momentum j, excitation energy $\epsilon_{k,l}$ (respect to threshold energy), with discretization step $\Delta\epsilon_l$ up to a maximum energy ϵ_{max}. The N_l bin-states are constructed in sequential intervals of energy, *i.e.*, $\epsilon_{i-1} < \epsilon_k < \epsilon_i$ with $i = 1, N_l$, where $\epsilon_0 = 0$ and $\epsilon_{Nlmax} = \varepsilon_{max}$. Within the CDCC model, bin states $\phi^{(i)}_{l,s,j}(r)$, $(i = 1, N_l)$ are square integrable and are constructed as a superposition of scattering wave functions within the interval $[k_{i-1}, k_i]$ with excitation energy corresponding to the mean value energy,

$$\varepsilon_i = \frac{\varepsilon_{k_i} + \varepsilon_{k_{i-1}}}{2}, \tag{1}$$

where,

$$\phi^{(i)}_{l,s,j}(r) = \frac{1}{\eta_i^{1/2}} \int_{k_{i-1}}^{k_i} w_i(k)\psi^{(scatt.)}_{l,s,j}(r,k)dk, \tag{2}$$

where $w_i(k)$ is a weight function that satisfies, $\int_{k_{i-1}}^{k_i} |w_i(k)|^2 dk = \eta_i$. Resonance energies $\epsilon_{i,res}$ and widths Γ_{res} of ^{6}Li correspond to the states $j^\pi = 3+, 2+$and $1+$ with values $\varepsilon_{3+} = 0.716$ MeV, $\Gamma_{3+} = 0.024$ MeV, $\varepsilon_{2+} = 2.84$ MeV, $\Gamma_{2+} = 1.7$ MeV and $\varepsilon_{1+} = 4.18$ MeV, $\Gamma_{1+} = 1.5$ MeV, respectively. Within the CDCC formalism, a set of coupled channel equations is obtained for the radial part of the relative motion wave function,

$$\left[T_R + U^{(J)}_{\beta\beta}(R) - E - \varepsilon_\beta\right] u^{(J)}_\beta(R) = -\sum_{\beta'} U^{(J)}_{\beta\beta'}(R)u^{(J)}_{\beta'}(R), \tag{3}$$

where angular coordinates have been integrated. Here, $\beta = l, L, J$ are the quantum numbers of the ground and discrete states, ε_β is the eigenvalue of the internal Hamiltonian of the nuclear system. $U^{(J)}_{\beta\beta}$ represents the diagonal matrix elements of the interactions and $U^{(J)}_{\beta\beta'}$ non-diagonal elements. In particular, the diagonal matrix element $U^{(J)}_{00}(R)$ corresponds to the optical potential in the elastic channel. Inelastic excitations of the target are not explicitely considered in our calculations, while all continuum couplings are included. The deuteron-target potential $V_{dT}(r)$ is that taken from Ref.[20].

Regarding the fragment-target interactions, the deuteron-target potential $V_{dT}(r)$ is that taken from Ref.[20]. This potential is the result of a large systematic study of existing experimental data of elastic scattering angular distributions and non-elastic cross sections of incident deuteron on a wide range of target nuclei and energies. It is given in terms of Woods-Saxon shapes as,

$$V(r) = -V_0 f_r(r) - iW_V f_V(r) + 4ia_s W_s \frac{df_S}{dr} + \lambda_\pi^2 \frac{V_{so} + iW_{so}}{r}\frac{df so}{dr}\sigma \bullet \mathbf{l} + V_{Coul}(r), \tag{4}$$

where $f_j(r) = 1/[1 + \exp(r - R_j)/a_j]$, $R_j = r_j A^{1/3}$ with $j = r, V, S, so$. All potentials strengths, diffuseness and radii are given by the systematics, in terms of the target mass and deuteron energy E_d. The non-local, double-folding density dependent interaction SPP has been used for the $\alpha + target$ sub-system.

$$V_{\alpha T}(r, E) = [N_R(E) + iN_I(E)]V_F(r)e^{-v(E)^2/c^2}, \tag{5}$$

r being the separation between the α-particle and target and $v(E)$ their relative velocity. The imaginary part is assumed with the same radial dependence as the real part. $N_R(E)$ and $N_I(E)$ are normalization coefficients that take into account polarization effects arising from direct reaction couplings and $V_F(r)$ is the folding potential given by,

$$V_F(r) = \int \rho_{A1}(\mathbf{r_1})\rho_{A2}(\mathbf{r_2})V_0\delta[\mathbf{r} - (\mathbf{r_1} - \mathbf{r_2})]d\mathbf{r_1}d\mathbf{r_2}, \tag{6}$$

with Fermi density distributions,

$$\rho_i(r) = \frac{\rho_{0i}}{1 + \exp\frac{r - R_i}{a_i}}.i = A_1, A_2. \tag{7}$$

Through an extensive systematic analysis of elastic scattering differential cross sections for several systems over a wide range of bombarding energies, global values have been determined for the zero- range potential strength $V_0 = -456$ MeV-fm^3, the mass density diffuseness $a_i = 0.56$ fm and the radius parameter $R_i = 1.31A_i^{1/3} - 0.84$ fm, $i = A_1, A_2$. Thus the SPP is a parameter-free interaction which depends only on the masses A_1, A_2 of the interacting nuclei and ther collision energy E_{lab}.

3. CDCC calculations of elastic scattering using global interactions

In this section CDCC calculations of elastic scattering angular distributions are presented for the projectile ^{6}Li with targets ^{144}Sm and ^{208}Pb for incident energies just above the corresponding Coulomb barriers. In our calculations, the code FRESCO is used[27], where regarding the V_{dT} interaction, care is taken for the radial parameters R_j. This is so, since in the code FRESCO, the general relation $R_j = r_j(A_1^{1/3} + A_2^{1/3})$ is considered while a shorter $R_j = r_j A^{1/3}$ is used in Ref.[20], being $A_1 = 2$ and A the target mass. To calculate the ground $\psi_{l=0}^{(0)}(E_{thre} = -1.47$ MeV), resonant $l = 2$, $j^\pi = 3^+, 2^+, 1^+$ and non-resonant breakup discrete states of ^{6}Li, we have used the parametrization for the interaction $V_{\alpha d}(r)$ given in Ref.[26]. The discretization is made as follows; the maximum angular momentum for the relative motion of the fragments $\alpha - d$ of ^{6}Li is $l_{max} = 3$. Larger values do not have any effect on the calculations. So, bin states are constructed for $l = 0$, $j^\pi = 1^+$ and $l = 1$, $j^\pi = 0^-, 1^-, 2^-$, with step $\Delta\varepsilon = 0.5$ MeV up to $\varepsilon_{\max} = 6.8$ MeV. Finer and variable steps are used for resonant states $l = 2$, $j^\pi = 3^+, 2^+, 1^+$, so as to obtain centroid excitation energies and widths close to the corresponding measured values. Table 1 shows the calculated values with the interaction $V_{\alpha d}$ of Ref.[26].

Table.1. Calculated values (MeV) for resonance energies and widths of ^{6}Li

Res	E_{cal}	Γ_{cal}	$E_{\exp}$	$\Gamma_{\exp}$
3^+	0.73	0.034	0.716	0.024
2^+	3.09	1.3	2.84	1.7
1^+	4.67	4.2	4.18	1.5

For bin states $l = 3$, $j^\pi = 4^-, 3^-, 2^-$, a larger step $\Delta\varepsilon = 1.0$ MeV is used. Convergence tests at $\varepsilon_{\max} = 7.0, 7.5$ and 8.0 MeV were done with no effect on elastic angular distributions. Similarly, larger steps $\Delta\varepsilon = 0.75$ and 1.0 MeV were used with no appreciable effect on elastic scattering. Partial waves for the relative motion of ^{6}Li and target are considered up to $L_{max} = 600$ and nuclear and Coulomb potential multipoles are included up to $L_Q = 4$ for which convergence is achieved. As regards the normalization coefficients of the interaction $V_{\alpha T}$ of Eq.(5), we set these as $N_R = 1$ and $N_I = 1$. This is so, since for the nuclear systems and range of collision energies studied in this work, the $\alpha - target$ relative kinetic energy in the center of mass system is, in most cases, above the Coulomb barrier. Hence, polarization effects are negligible. The effect of resonances of ^{6}Li ($l = 2, j^\pi = 3+, 2+, 1+$) on elastic scattering differential cross sections is studied by removing the narrow range of continuum bins around themselves. Figs.1 and 2 show the results for the elastic scattering angular distributions for the systems ^{6}Li+^{144}Sm and ^{6}Li+^{208}Pb for laboratory energies just above the barrier. As seen in Fig.1, for ^{6}Li+^{144}Sm close agreement with the data is achieved when all couplings are considered (solid-lines). For the case when resonant continuum states are removed, the effect on elastic cattering is very small, except at the lowest energies (dashed-lines). In this case the calculations are slightly below those with resonance space included indicating a weak attractive polarization potential contribution from these states. At all energies, it is evident the importance of couplings to continuum states, since the calculations with only coupling to the elastic channel, that is, considering only the term U_{00}^J in Eq.(3) is well below the data (dashed-lines). As for ^{208}Pb, Fig.2 shows that except for the lowest energy the calculations with couplings to all continuum states are in agreement with the data. The effect of couplngs to resonance states of ^{6}Li is in most cases small except at $E_{lab} = 39$ Mev at large angles.

In summary, CDCC calculations of the weakly bound projectile ^{6}Li with targets ^{144}Sm and ^{208}Pb have been performed with global fragment-target interactions dependent on the target mass and incident energy. In most cases, the results for the elastic scattering angular distributions at energies just above the barrier closely agree with the data when couplings to

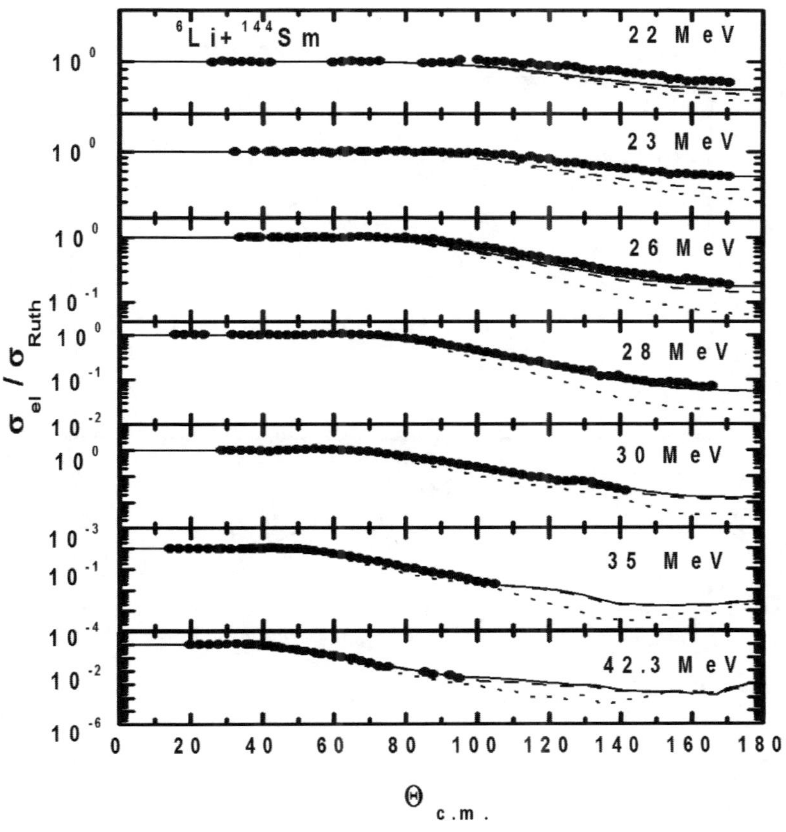

Figure 1. CDCC calculations for the elastic scattering of ^{6}Li+^{144}Sm at the laboratory energies shown. Solid-lines correspond to the calculations with all couplings as descibed in the text. Dashed-lines represent the calculations with resonance states of ^{6}Li extracted from the continuum. Dotted-lines correspond to single elastic channel calculations without couplings to continuum states. Data are taken from Refs.[28]

continuum states of ^{6}Li are considered. Optical model calculations, without continuum couplings predict elastic scattering angular distributions well below the experimental values. This could mean that continuum couplings to breakup channels of ^{6}Li produce a net repulsive polarization potential that increases the Coulomb barrier. This fact, consequently leads to less absorption from the elastic channel. On the same footing, the effect of continuum resonant states of ^{6}Li on elastic scattering has been studied. The effect appears to be not too strong in most cases.

4. Acknowledgments

The author acknowledges the support from CONACYT, Mexico.

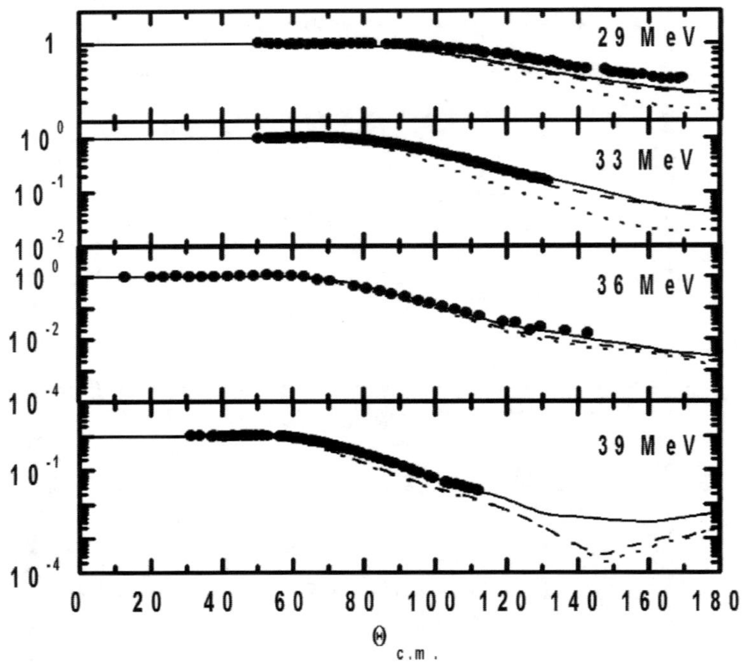

Figure 2. Same as Fig.1 but for $^{6}Li+^{208}Pb$. Data are taken from Refs.[29, 30, 31]

References

[1] Canto L F, Gomes P R S, Donangelo R, Hussein M S, 2006 *Phys. Rep.* **424** 1
[2] Liang J F, Signorini C, 2005 *Int. J. Mod. Phys.* E **14** 1121
[3] Keeley N, Raabe R, Alamanos N, Sida J L, 2007 *Prog. Part. Nucl. Sci.* **59** 579
[4] Beck C, Keeley N, Diaz-Torres A, 2007 *Phys. Rev.* C **75** 054605
[5] Fernandez-Garcia J P, *et al.*, 2010 *Phys. Lett.* B **693** 310
[6] Sakuragi Y, Yahiro M, Kamimura M, 1986 *Prog. Theor. Phys.* Suppl. **89** 1
[7] Zerva K *et al.*, 2009 Phys. Rev. C **80** 017601
[8] Rusek K, Keeley N, Pakou A, Alamanos A, 2007 *Nucl. Phys.* A **784** 13
[9] Pakou A, *et al.*, 2006 *Phys. Lett.* B **633** 691
[10] Keeley N, Mackintosh R S , Beck C, 2010 *Nucl. Phys.* A **380** 1
[11] Keeley N, Rusek K, 1996 *Phys. Lett.* B **375** 9
[12] Otomar D R, Lubian J, Gomes P R S, 2010 *Eur. Phys. J.* A. **46** 285
[13] Keeley G R, *et al.*, 2000 *Phys. Rev.* C **63** 024601
[14] Pakou A, *et al.*, 2004 *Phys. Rev.* C **69** 054602
[15] Pakou A *et al.*, 2009 *Eur. Phys. J* A **39** 187
[16] Otomar D R, *et al.* 2009 *Phys. Rev.* C **80** 034614
[17] Canto L F, Lubian J, Gomes P R S, Hussein M S, 2009 *Phys. Rev.* C **80** 047601
[18] Lubian J, Correa T, Aguilera E F, Canto L F, Gomez Camacho A, Quiroz E M, Gomes P R S, 2009 *Phys. Rev.* C **79** 064605
[19] Austern N, Iseri Y, Kamimura M, Kawai M, Rawitscher G, Yahiro M, 1987 *Phys. Rep.* **154** 125
[20] Haixia A, Chonghai C, 2006 *Phys. Rev.* C **73** 054605
[21] Perey C M, Perey F G, 1974 *At. Data Nucl. Data Tables* **13** 293; 1974 *At. Data Nucl. Data Tables* **17** 1
[22] Daehnick W W, Childs J D, Vrcelj Z, 1980 *Phys. Rev.* C **21** 2253
[23] Bojowald J, Machner H, Nann H, Oelert W, Rogge M, Turek P, 1988 *Phys. Rev.* C **38** 1153
[24] Chamon L C, *et al.*, 1997 *Phys. Rev. Lett.* **79** 5218

[25] Chamon L C, *et al*, *Phys. Rev.* C **66** 014610
[26] Diaz-Torres A, Thompson I J, Beck C, 2003 *Phys. Rev.* C **68** 044607
[27] Thompson I J, 1988 *Comput. Phys. Rep.* **7** 167
[28] Figueira J M, *et al.*, 2010 Phys. Rev. C **81** 024613
[29] Keeley N, Bennett S J, Clarke N M, Fulton B R, Tungatc G, Drumm P V, Nagarajan M A, Lilley J S, 1994 *Nucl. Phys.* A **571** 326
[30] Gemmeke H, Deluigi B, Lassen L, Scholz D, 1978 *Z. Physik* A **286** 73
[31] Cook J, Gils H J, Rebel H, Majka Z, Klewe-Nebenius H, 1982 *Nucl. Phys.* A **388** 173

Journal of Physics: Conference Series **578** (2015) 012004 doi:10.1088/1742-6596/578/1/012004

Identified particle production in pp, p–Pb, and Pb–Pb collisions measured with ALICE at LHC energies.

Raúl Tonatiuh Jiménez-Bustamante (on behalf of the ALICE Collaboration)

Instituto de Ciencias Nucleares, Universidad Nacional Autónoma de México. Apartado Postal 70-543, México Distrito Federal 04510, México.

E-mail: raul@cern.ch

Abstract. The ALICE detector has excellent Particle IDentification (PID) capabilities in the central barrel ($|\eta| < 0.9$). This allows identified hadron production to be measured over a wide transverse momentum ($p_{\rm T}$) range, using different sub-detectors and techniques: their specific energy loss ($\mathrm{d}E/\mathrm{d}x$), the time of flight, the Cherenkov angle or their characteristic weak decay topology. Results on identified particle spectra and production yield ratios at mid-rapidity measured by ALICE in different colliding systems (pp, p–Pb and Pb–Pb) are presented and the similarities among them are discussed. For Pb–Pb collisions the nuclear modification factor as a function of $p_{\rm T}$ is shown for different collision centralities.

1. Introduction

ALICE is a general-purpose, heavy-ion detector which focuses on QCD, the strong-interaction sector of the Standard Model. It is designed to address the physics of strongly interacting matter and the quark-gluon plasma at extreme values of energy density and temperature in nucleus-nucleus collisions. Detailed description of the ALICE experiment and its detector subsystems can be found in [1], [2] and [3]. The heavy ion collisions produce a system of hot dense matter that expands due to the pressure gradients between the system center and the surrounding vacuum [4]. The system expands collectively and as a consequence it cools down, and undergoes hadronization. The measurement of the identified transverse momentum ($p_{\rm T}$) spectra provides information about the hadronization processes and the evolution of the created system. In Pb–Pb collisions the low transverse momentum region ($p_{\rm T} < 3$ GeV/c) provides information about the collective properties of the system like radial flow, where the $p_{\rm T}$ spectra have a strong dependence on the mass of the particle. The intermediate momentum region (3 GeV/$c < p_{\rm T} < 8$ GeV/c) provides information about hadronization mechanisms according to models like quark-recombination [5], where the relevant variable is the number of constituent quarks. In the high momentum region ($p_{\rm T} > 8$ GeV/c) the effects in the fragmentation due to the medium created can be explored [6].

Information from pp collisions are important not only as a baseline for the Pb–Pb measurements, but also because they give the possibility to improve Monte Carlo generators based on QCD models. The p–Pb information is used to evaluate cold nuclear effects. Recent

ALICE results suggests that p–Pb collisions present effects similar to those observed in Pb–Pb collisions like flow [7].

Previous results on identified particle production in pp, p–Pb and Pb–Pb can be found in [7], [8] and [9]. ALICE results on the nuclear modification factor have been reported in [10].

2. Results

Fig. 1 (left) shows the p_T distributions of charged pions, kaons and protons in pp collisions at $\sqrt{s} = 2.76$ TeV obtained with the PID information from the Inner Tracking System (ITS), the Time Projection Chamber (TPC), the Time Of Flight detector (TOF) and the High Momentum Particle Identification Detector (HMPID). The combined information from these detectors allow us to measure the spectra within the rapidity window $\mid y \mid < 0.5$ in the p_T range $0.1 - 20$ GeV/c.

In Pb–Pb collisions the analysis is performed in different centrality classes which are both determined with the amplitude measured in the VZERO detectors and simulations based on the Glauber model [11]. In p–Pb collisions the spectra are reported in multiplicity classes due to the weak correlation between collision geometry and the multiplicity based on the amplitude of the signal of VZEROA detector which is located in the Pb side [7].

The p_T spectra of pions, kaons and protons measured by ALICE in the most central Pb–Pb collisions ($0 - 5$ %) are shown in Fig. 1 (right). The p_T spectra measured by ALICE are flatter than those measured at RHIC energies specially for protons. This indicates a stronger radial flow when increasing the energy. The comparison to models based on hydrodynamics show that VISH 2+1 [12] describes the pion and kaon spectra for $p_T < 1.5$ GeV/c qualitatively well, but fails in the description of protons, which can be attributed to a lack of an explicit description of the hadronic phase. HKM [13] inserts a hadronic cascade after the hydrodynamic phase, which further transports them until final decoupling. This improves the description of the protons for $p_T < 1.0$ GeV/c. The Kraków [14] model provides a better description of the spectra than the other 2 models. This model uses changes in the effective temperature (T_{ch}), consequence of non-equilibrium due to the bulk viscosity. The EPOS [15] model uses breakup of flux tubes which either contribute to the bulk or escape the medium as jets. This model describes qualitatively well the data in a broad p_T range.

The ratio $(\mathrm{p}+\bar{\mathrm{p}})/(\pi^+ + \pi^-) = \mathrm{p}/\pi$ is shown in Fig. 2 for two different centralities in Pb–Pb collisions compared to pp collisions. The central events show an enhancement in the intermediate p_T region ($3 - 8$ GeV/c), compatible with previous observations at RHIC in Au–Au collisions at $\sqrt{s_{NN}} = 200$ GeV [16], [17]. ALICE has also measured the ratio $(\mathrm{K}^+ + \mathrm{K}^-)/(\pi^+ + \pi^-) = \mathrm{K}/\pi$ and (p/π) at different pp energies [9] and found no significant energy dependence within the experimental errors.

The final state effects in Pb–Pb collisions can be studied by the nuclear modification factor (R_{AA}). The R_{AA} is defined as:

$$R_{AA} = \frac{\mathrm{d}^2 N_{ch}^{PbPb}/\mathrm{d}\eta \mathrm{d}p_T}{\langle N_{coll} \rangle \mathrm{d}^2 N_{ch}^{pp}/\mathrm{d}\eta \mathrm{d}p_T} \quad (1)$$

where $\langle N_{coll} \rangle$ is the number of binary collisions for different centrality classes [11].

Fig. 3 shows that at $p_T > 8$ GeV/c the suppression for pions, kaons and protons is the same within systematic uncertainties, for central and peripheral Pb–Pb collisions. This measurement disfavors significant modifications of hadro-chemistry within the hard core of jets [6].

ALICE has shown that in p–Pb collisions, the nuclear modification factor, R_{pPb}, for charged particles is consistent with one for $p_T \geq 2$ GeV/c [18]. This means no suppression is observed in p–Pb collisions unlike in Pb–Pb collisions. This indicates that initial state effects are small. However, the transverse momentum spectra of pions, kaons and protons show similar behavior to that observed in Pb–Pb collisions associated with the collective expansion of a medium.

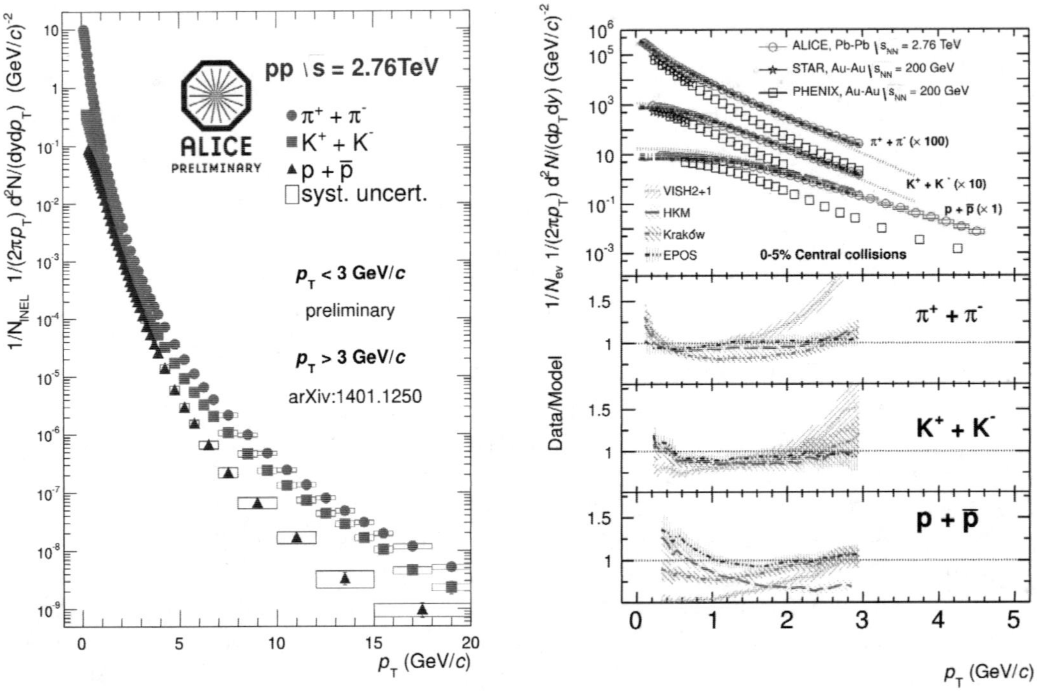

Figure 1. Left: π/ K/ p in pp collisions at $\sqrt{s}$ = 2.76 TeV. Right: Comparison of π/ K/ p spectra to previous observations in Au–Au collisions at $\sqrt{s_{NN}}$ = 200 GeV and hydro models in Pb–Pb central collisions (0 – 5%).

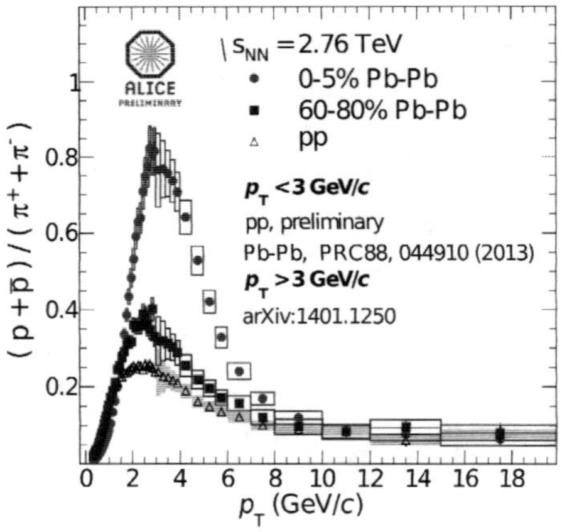

Figure 2. Proton to pion ratio measured in pp and Pb–Pb collisions.

The baryon to meson ratios: p/π and Λ/K^0_s for two extreme measured centrality (multiplicity) classes are shown in Fig. 4 for Pb–Pb (p–Pb) collisions. The behavior as a function of p_T for both systems is similar, showing an enhancement at intermediate p_T (at $\approx$ 3.0 GeV/c) for high centrality or multiplicity.

The collective expansion can be tested with a blast-wave fit simultaneous to all particles in a

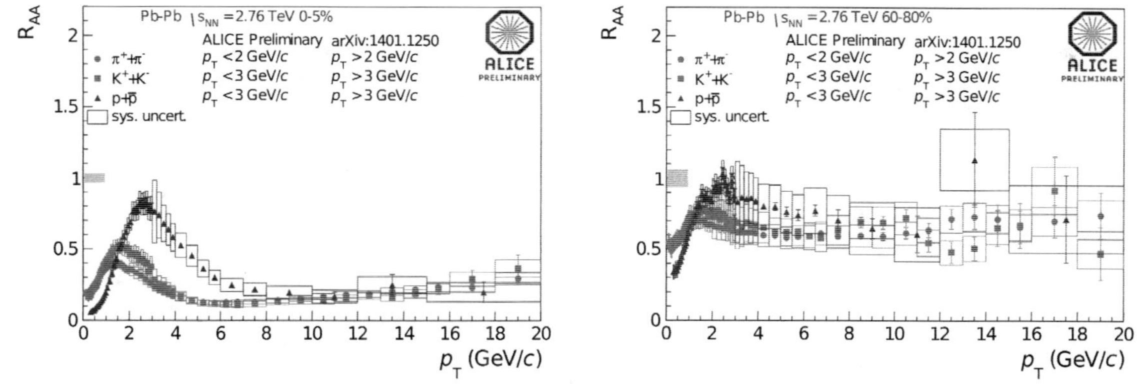

Figure 3. Nuclear modification factor for pions, kaons and protons, for central (left) and peripheral collisions.

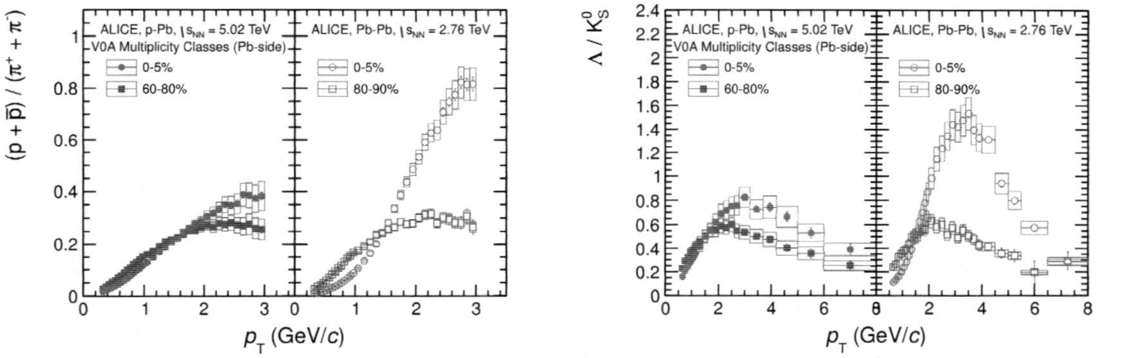

Figure 4. Baryon to meson ratio: p/π (left) and Λ/K^0_S (right) as a function of p_T measured in two multiplicity bins in p–Pb collisions compared to results for two centrality bins in Pb–Pb collisions.

multiplicity bin. This is a traditional way to compare the p_T distributions and their evolution in different systems. These fits to all particles species provide a kinetic freeze-out temperature T_{kin} and the expansion velocity $\langle\beta_T\rangle$ of the system. Fig. 5 shows the parameters obtained with the blast-wave fit for the 3 different systems: pp, p–Pb and Pb–Pb. The behavior is qualitatively similar in all of them, this observation indicates the presence of radial flow in the 3 systems. On the other hand, the results obtained with simulations based on PYTHIA 8 (tune 4C) [19], which do not include any collective effect, produce a similar pattern, because of a color reconnection mechanism mimicking flow [20].

3. Conclusions

ALICE measurements of pion, kaon and proton production as a function of p_T at several energies for different colliding systems have been presented. ALICE results in pp collisions show no energy dependence on the ratios p/π and K/π. In Pb–Pb collisions the p_T spectra are qualitatively well reproduced by hydrodynamical models. However one should bear in mind that all the models, even EPOS, fail in the quantitative description of the data in all the p_T range. The R_{AA} at high p_T seems to disfavor particle species (flavor) dependent effects. The baryon to meson ratios in p–Pb collisions exhibit similar enhancement to that observed in Pb–Pb collisions generally

Journal of Physics: Conference Series **578** (2015) 012004 doi:10.1088/1742-6596/578/1/012004

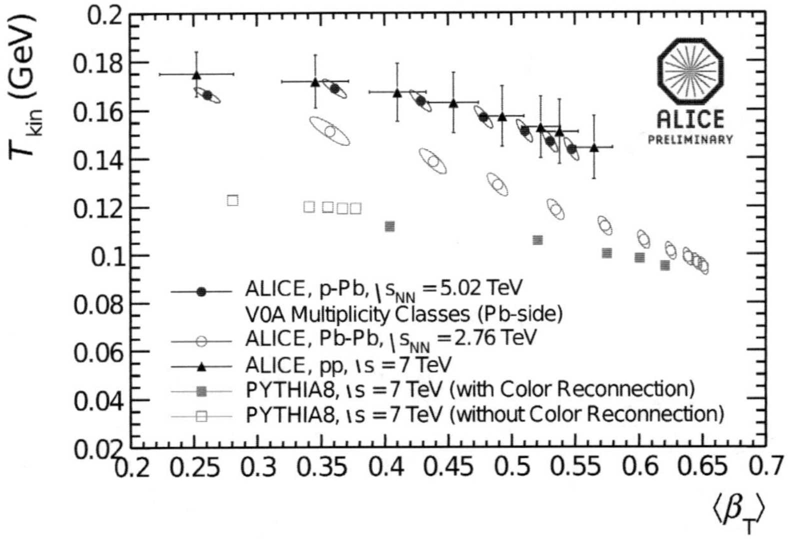

Figure 5. Blast-wave fit analysis, compared to Pb–Pb, p–Pb, pp data and MC simulations using PYTHIA8 with and without color reconnection. Charged-particle multiplicity increases from left to right.

related to collective flow. Blast wave fits in Pb–Pb, p–Pb and high multiplicity pp collisions show similar flow signatures, however this is observed also in simulations using PYTHIA 8, which has no hydrodynamics present, but instead has a color reconnection mechanism that mimics radial flow effects [20].

References

[1] ALICE Collaboration 2006 *J. Phys.* G **32** 1295
[2] ALICE Collaboration 2004 *J. Phys.* G **30** 1517
[3] ALICE Collaboration 2008 *J. Instrum.* **3** S08002.
[4] Heinz U W 2013 Towards the Little Bang Standard Model *Preprint* nucl-th/1304.3634
[5] Fries R J, Greco V and Sorensen P 2008 *Ann. Rev. Nucl. Part. Sci.* **58**, 177, 0807.4939
[6] Sapeta S and Wiedemann U A 2008 *Eur. Phys. J.* C **55** 293
[7] ALICE Collaboration 2014 *Physics Letters* B **728** 25
[8] ALICE Collaboration 2013 *Phys.Rev.* C **88** 044910
[9] Chojnacki M 2011 Measurement of π, K , p transverse momentum spectra with ALICE in proton-proton collisions at $\sqrt{s}$ = 0.9 and 7 TeV. *Preprint* hep-ex/1109.6744
[10] ALICE Collaboration 2014 Production of charged pions, kaons and protons at large transverse momenta in pp and Pb–Pb collisions at $\sqrt{s_{\mathrm{NN}}}$ = 2.76 TeV *Preprint* nucl-ex/1401.1250
[11] ALICE Collaboration 2013 *Phys.Rev.* C **88** 044909
[12] Shen C, Heinz U , Huovinen P and Song H 2011 *Phys. Rev.* C **84** 044903
[13] Karpenko I, Sinyukov Y and Werner K 2013 *Phys. Rev.* C **87** 024914
[14] Bozek P 2012 *Phys. Rev.* C **85** 034901
[15] Werner K, Karpenko I, Bleicher M, Pierog T and Porteboeuf-Houssais S 2012 *Phys. Rev.* C **85** 064907
[16] PHENIX Collaboration 2013 Spectra and ratios of identified particles in Au+Au and d+Au collisions at $\sqrt{s_{NN}}$=200 GeV *Preprint* nucl-ex/1304.3410
[17] STAR Collaboration 2006 *Phys. Rev. Lett.* **97** 152301
[18] ALICE Collaboration 2013 *Phys. Rev. Lett.* **110** 082302
[19] Corke R and Sjostrand T 2011 *JHEP* **03** 32
[20] Ortiz A, Christiansen P, Cuautle E, Maldonado I and Paic G 2013 *Phys. Rev. Lett.* **111** 042001

Journal of Physics: Conference Series **578** (2015) 012005 doi:10.1088/1742-6596/578/1/012005

Study of the PYTHIA $\Delta\eta\Delta\phi$ correlation as a function of multiplicity and transverse sphericity in proton-proton collisions at 7 TeV.

E Pérez Lezama, G Paić, E Cuautle Flores, A Ortiz Velasquez

Instituto de Ciencias Nucleares, Universidad Nacional Autónoma de México Apartado Postal 70-543, 04510 México, D. F., México

E-mail: edgar.perez.lezama@cern.ch

Abstract. The study and understanding of the details of proton-proton collisions is important since they are the benchmark for the comparison with heavy ion results. In the present work we have studied the $\Delta\eta\Delta\phi$ correlations as a function of the event multiplicity and transverse sphericity in pp collisions at $\sqrt{s} = 7$ TeV using the PYTHIA 8.180 generator. The results show large variation in the shape of the correlation as function of the sphericity and multiplicity indicating the importance of slicing the data in bins of both observables to better understand the complexity of the interactions. Motivated by the flow like patterns originated by color reconnection (CR), results with and without CR together with dihadron correlations are discussed.

1. Introduction

The pp collisions have a complicated structure which involves complex final states, with large multiplicities of hadrons, all originated from different processes [1], this complexity of the interactions requires MC event generators to simulate them . PYTHIA[2] is an event generator which comprises a coherent set of physics models for the evolution from a few-body hard process to a complex multihadronic final state. Hadronic processes can be classified as being either soft or hard.

Different techniques of analysis are used to investigate the properties of the hadronic interactions, among them the dihadron azimuthal correlations. Using this observable, an interesting ridge-shaped correlation has been discovered in heavy nuclei collisions. Some models attribute the ridge to jet-medium interactions, while others attribute it to the medium itself [3]-[4]. Recent results from CMS have observed a striking ridge structure in very high multiplicity proton-proton (pp) collisions at a center-of-mass energy of 7 TeV. This novel structure resembles similar features observed in relativistic heavy-ion experiments [3]. Several mechanisms like Color-Glass Condensate based on initial state nonlinear gluon interactions[5] and elliptic flow[6] have been proposed to explain the phenomenon.

Although the effect observed by CMS, cannot be reproduced[4] by PYTHIA 8.180, in this work we investigate the possible presence of the ridge, studying the dihadron correlation as a function of the event shape variable, since it allows to control the amount of MPI[7]. According with Ref.[8] the color reconnection mechanism (CR) produces effects which mimics radial flow

Journal of Physics: Conference Series **578** (2015) 012005 doi:10.1088/1742-6596/578/1/012005

in events with a large number of multi-partonic-interactions and the same effect might produce elliptic flow.

At hadron colliders the event shape analyses are restricted to the transverse plane in order to avoid the bias from the boost along the beam axis. The transverse sphericity (S_{T}) is defined in terms of eigenvalues $\lambda_1 > \lambda_2$ of the transverse momentum matrix:

$$S_{xy} = \frac{1}{\sum_i p_{\mathrm{T_i}}} \sum_i \frac{1}{p_{\mathrm{T_i}}} \begin{pmatrix} p_{x_i}^2 & p_{x_i}p_{y_i} \\ p_{y_i}p_{x_i} & p_{y_i}^2 \end{pmatrix} \tag{1}$$

in the following way

$$S_{\mathrm{T}} \equiv \frac{2\lambda_2}{\lambda_2 + \lambda_1}. \tag{2}$$

The limit values of this variable are related to specific configuration in the transverse plane

$$S_{\mathrm{T}} = \begin{cases} 0 & \text{"pencil like" limit} \\ 1 & \text{"isotropic" limit} \end{cases}$$

2. Analysis Details

Using the PYTHIA 8.180 event generator we analyzed primary charged particles from 500 million of events with CR and the same without CR. In order to have a good description of the sphericity, events with more than two particles and $p_{\mathrm{T}} > 0.5$ GeV/c are selected, and is calculated with all the particles within the pseudorapidity region $|\eta| < 1.0$. The leading particle is selected in the pseudo-rapidity range $| \eta |< 1.0$, the same as sphericity, and having the η range open for the associated particle. In this work the trigger and associated particle are not selected in a certain momentum range, as done in the conventional dihadron correlation analyses. Our leading particle is the one with the highest transverse momentum in each event, and the associated particles are all the others that fulfil the p_{T} and η cuts previously described. Thus, the correlation is built by taking the differences in η and ϕ between the leading and the associated particles:

$$\begin{aligned} \Delta\eta &= \eta^{assoc} - \eta^{lea} \\ \Delta\phi &= \phi^{assoc} - \phi^{lea} \end{aligned} \tag{3}$$

Where η^{lea} and ϕ^{lea} (η^{assoc}, ϕ^{assoc}) are the pseudo-rapidity and the azimuthal angle of the leading particle (associated), respectively.

A plot of the dihadron correlations for minimum bias events is shown in figure 1. Two prominent structures are visible, one at $(\Delta\phi, \Delta\eta) \approx 0$, known as "near side peak", and correspond to correlations from single jets. The second structure the"away side peak" is present at $\Delta\phi \approx \pi$ and corresponds to back-to-back jet correlations.

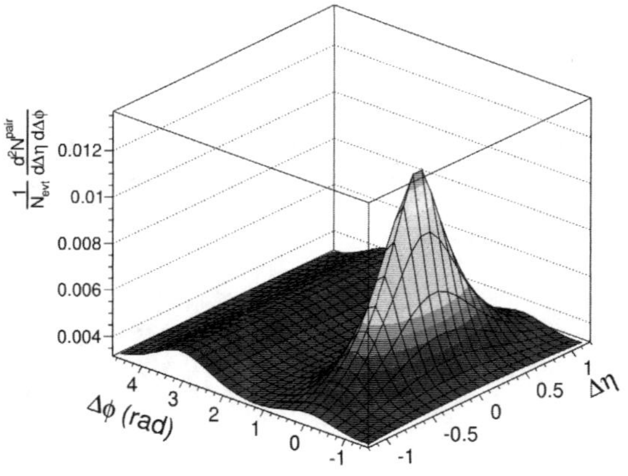

Figure 1. $\Delta\eta - \Delta\phi$ correlation for a leading and associated particle selected in the region $|\eta| < 1.0$, where multiplicity and sphericity cuts are not applied.

3. Results

The results are presented using two sphericity ranges: $S_T < 0.1$ and $0.9 < S_T$, jetty-like and isotropic events, respectively. The correlations are computed as a function of multiplicity of the events. The multiplicity is expressed in function of $z = (dN_{ch}/d\eta)/\langle dN_{ch}/d\eta\rangle$, i.e. the ratio of the multiplicity and the mean multiplicity. The multiplicity ranges presented are: $z < 1$, $2 < z < 3$ and $4 < z$. The correlations are normalized to the number of leading particles(same as number of events) in each sphericity range being analyzed.

Figure 2 shows the dihadron correlation for low sphericity events (jetty-like,$S_{\mathrm{T}} < 0.1$), where no ridge proper structure has been observed in the ranges studied although an extended structure of the near side peak is observed that will be the subject of farther studies.

Journal of Physics: Conference Series **578** (2015) 012005 doi:10.1088/1742-6596/578/1/012005

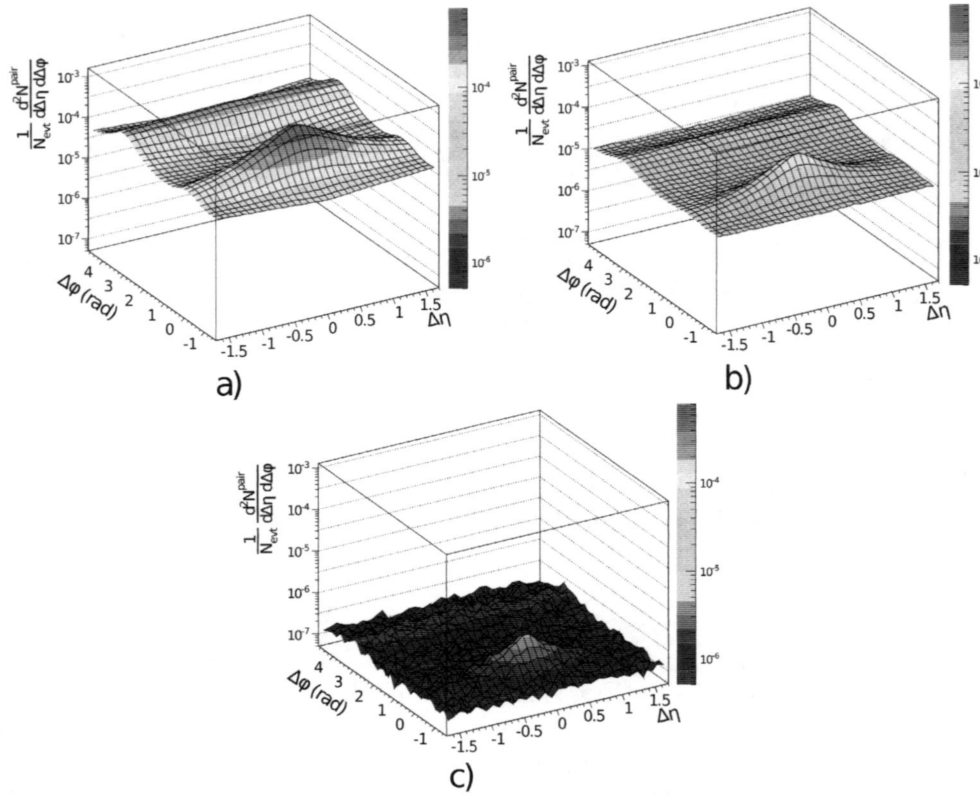

Figure 2. Dihadron correlations for low sphericity events ($S_{\rm T} < 0.1$) in three multiplicity ranges, CR is activated.(a) for $z < 1$; (b) for $2 < z < 3$ and (c) for $4 < z$.

Figure 3 shows the dihadron correlations for $S_{\rm T} > 0.9$ in three multiplicity ranges, where an interesting and unexpected double hump structure appears, even though in PYTHIA the processes $2 \rightarrow 3$, that are observed as "T" event structures, are suppressed with respect to the $2 \rightarrow 2$ processes. Similar structures are present when color reconnection mechanism is switched off, so the shape is not due to a flow effect. This has been already reported in [9].

The projections on the $\Delta\phi$ axis of the dihadron correlations, normalized to the number of events in the multiplicity range, show the underlying structures for three different ranges in the transverse sphericity. In the jetty events ($S_{\rm T} < 0.1$) we observe the expected near and away side peaks, for high sphericity ($0.9 < S_{\rm T}$) we observe the double hump . For the intermediate sphericity ($0.4 < S_T < 0.7$) we observe a broad away side with some structure. Figure 4 shows these projections together with the black curve representing the projection of all events in a given multiplicity range. We can see that the total correlation shape is a sum of different structures, that are identified by the use of the sphericity variable.

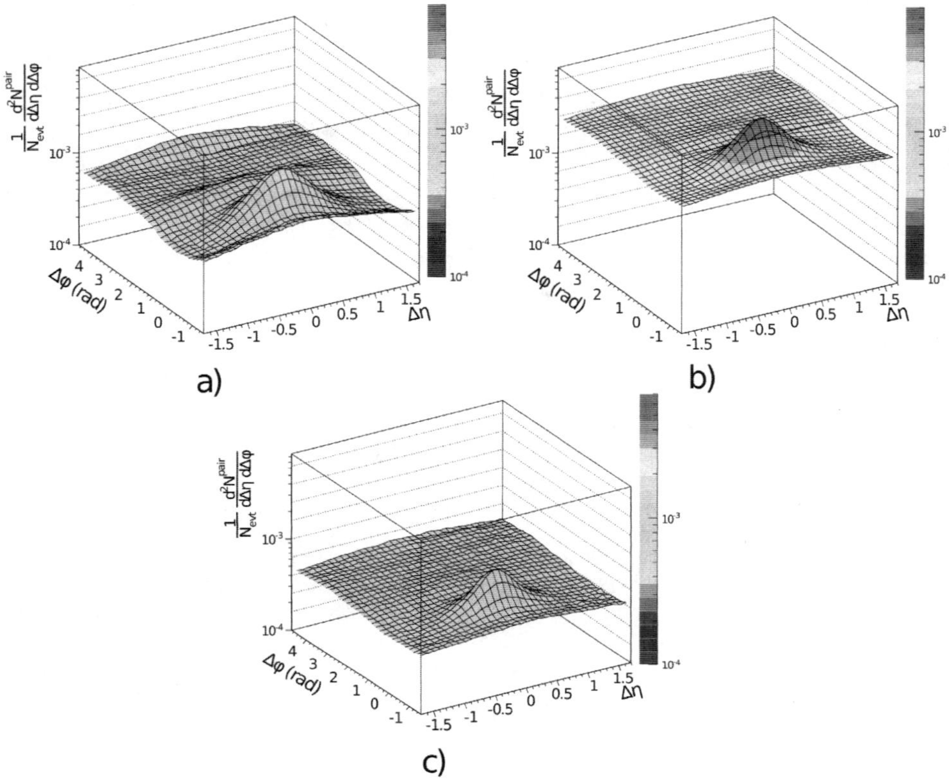

Figure 3. Dihadron correlations for high sphericity events ($S_T > 0.9$) in three multiplicity ranges, CR is activated, (a) for $z < 1$; (b) for $2 < z < 3$ and (c) for $4 < z$.

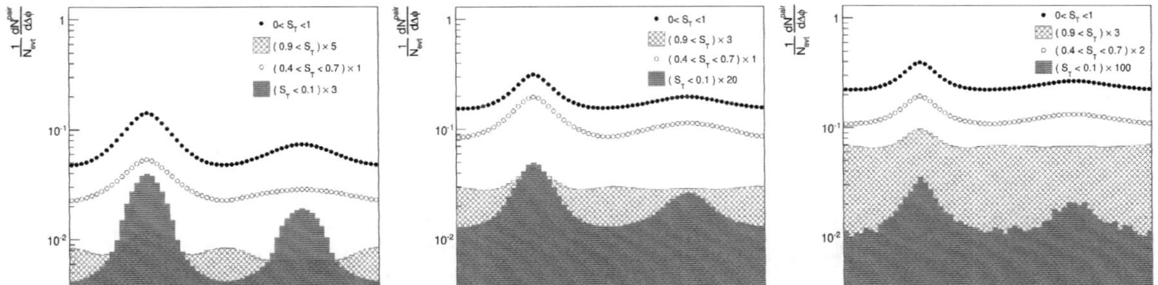

Figure 4. Projection in the $\Delta\phi$ axis selecting three S_T regions and three multiplicity ranges: (a) for $z < 1$; (b) for $2 < z < 3$ and (c) for $4 < z$. Black points represent the events with no sphericity cut.

4. Conclusions

In this work we have introduced transverse sphericity to analyze dihadron correlations. By using this event shape variable, we were able to analyze different types of events like those dominated by jetty like and/or isotropic structures. For high sphericity events we observe that the "T" structures events are enhanced, it is seen in the dihadron correlation figure 3 where double hump structures are present, this structures are located at $\Delta\phi \approx \pi/2$ and $\Delta\phi \approx -\pi/2$. We have shown that by sectioning p-p events in ranges of sphericity one uncovers the so called underlying event[10], usually treated as an amorphous structure. Actually has a rich structure and the usual treatment consisting to study the prominent jetty like structures in the near and away side subtracting a uniform background using the so called ZYAM method may give erroneous results.

Acknowledgments

Partial support for this work has been received by CONACyT under the grant numbers 103735 and 101597; and PAPIIT-UNAM under the projects: IN105113 and IN108414.

References

[1] Aamodt K *et al* (ALICE collaboration) 2010 *Eur. Phys. J.* C **68** 345
[2] Sjostrand T, Eden P, Friberg C, Lonnblad L, Miu G, Mrenna S and Norrbin E, 2001 *Computer Physics Commun.* **135** 238.
[3] Abelev B I *et al* (STAR collaboration) 2009 *Phys. Rev.* C **80** 064912
[4] Velicanu D 2011 *J. Phys.* G **38** 124051
[5] Dusling K and Venugopalan R 2013 *Phys. Rev.* D **87** 094034
[6] Werner K, Karpenko I and Pierog T 2011 *Phys. Rev. Lett.* **106** 122004
[7] Cuautle E *et al* 2014 Disentangling the soft and hard components of the pp collisions using the sphero(i)city approach *Preprint* arXiv:1404.2372
[8] Sjöstrand T, Mrenna S and Skands P 2006 *JHEP* **05(2006)** 026
[9] Ayala A *et al* 2009 *Eur. Phys. J.* C **62** 535 (*Preprint* arXiv:0902.0074)
[10] Field R 2012 *Acta Phys. Polon.* B **42** 2631 (*Preprint* arXiv:1202.0901)

Journal of Physics: Conference Series **578** (2015) 012006 doi:10.1088/1742-6596/578/1/012006

On the definition of entropy for quantum unstable states

Osvaldo Civitarese [1] **and Manuel Gadella** [2]

[1] Department of Physics, University of La Plata, C.C.67 (1900) La Plata, Argentina
[2] Department of FTAO, Faculty of Science. University of Valladolid, 47011, Valladolid, Spain

E-mail: osvaldo.civitarese@fisica.unlp.edu.ar

Abstract. The concept of entropy is central to the formulation of the quantum statistical mechanics, and it is linked to the definition of the density operator and the associated probabilities of occupation of quantum states. The extension of this scheme to accommodate for quantum decaying states is conceptually difficult, because of the nature of these states. Here we present a way to treat quantum unstable states in the context of statistical mechanics. We focuss on the definition of the entropy and avoid the use of complex temperatures.

1. Introduction

The fact that decaying states can be accommodated in the context of extensions of the framework of ordinary quantum mechanics [1] suggests that, in principle, they can also be accommodated in the framework of quantum statistical mechanics. However, the subject has been rarely treated in the literature, perhaps because of the inherent difficulty of having to deal with complex energy states in a formalism strongly limited by the notion of probabilities. Thus, one encounters here the same conceptual difficulties of the quantum mechanical treatment of complex energy states, which, as said before, can be avoided by adopting the notion of analytic extensions of the representations [1, 2, 3]. Kobayashi and Shimbori [4], and Kobayashi [5]have elaborated on the notion of entropy for complex-energy systems. In these papers the real and imaginary parts of the energy of a resonance are treated independently, and the canonical partition function for the resonance is given as a product of canonical partition functions for the real and imaginary parts of the energy, so that the total entropy is the sum of both contributions. Then, decaying processes transfer entropy from the imaginary part to the real part and the rate of this transference depends on time. Each part has its own temperature, which suggest a notion of a complex temperature. Another approach has been advanced by the Brussels group [6], by the notion of entropy operator for Gamow states, in the spirit of the entropy operator of Misra, Prigogine and Courbage [6]. This operator belongs to a family of operators defined on an extended space. The difficulty associated to this approach consists on the construction of the entropy operator.

In our approach, we give an expression for the entropy, without the introduction of the entropy operator of Misra et al.[6]. We assume thermodynamical equilibrium for decaying systems, which is a valid approximation if the half life of the state is sufficiently large, or equivalently if the width is sufficiently small, and use the Friedrichs model [7] as a laboratory, because it has the advantage of being a solvable model for resonance phenomena. We have formulated the Friedrichs model in second quantization language to construct coherent state and used of path integral method

Journal of Physics: Conference Series **578** (2015) 012006 doi:10.1088/1742-6596/578/1/012006

to obtain the canonical partition function of a system with resonances, which naturally leads to an expression for the canonical entropy of decaying states. The resulting entropy is complex, a fact that requires an interpretation. For the sake of the present work, and in order to introduce this method of calculation of the entropy for decaying systems, we have illustrated it to evaluate the canonical entropy for the harmonic oscillator. Although the path integration over coherent states yields a first approximation to the exact result, it shows its power to handle the problem of dealing with complex energy states. Further details of the formalism are presented in [8, 9]

2. The Friedrichs model in second quantification language.

The Friedrichs model [7], is an exactly solvable model for decaying phenomena in quantum mechanics, which shows all the features of resonant scattering. The point of departure is a unperturbed Hamiltonian H_0 with a simple absolutely continuous spectrum, and a positive eigenvalue, i.e., imbedded in the continuous spectrum. H_0 is written

$$H_0 = \omega_0\, a^\dagger a + \int_0^\infty d\omega\, \omega\, b_\omega^\dagger b_\omega\,, \tag{1}$$

where $a^\dagger$(a) is the creation (annihilation) operator of a bound state of energy ω_0, and $b_\omega^\dagger$ (b_ω) is the creation (annihilation) operator for the state of energy ω. The potential V is given by

$$V = \int_0^\infty d\omega\, f(\omega)\, (a^\dagger b_\omega + a b_\omega^\dagger)\,. \tag{2}$$

The function $f(\omega)$ in (2) is the form factor. The total Hamiltonian is $H = H_0 + \lambda V$, where λ is a real coupling constant. Creation and annihilation operators fulfill the following commutation relations:

$$[a, a^\dagger] = 1\,; \qquad [b_\omega, b_{\omega'}^\dagger] = \delta(\omega - \omega') \tag{3}$$

and all other commutators vanish. For the states we use the notation

$$|1\rangle = a^\dagger|0\rangle\,; \qquad |\omega\rangle = b_\omega^\dagger|0\rangle\,. \tag{4}$$

where $|0\rangle$ is the vacuum. Note that

$$a|0\rangle = b_\omega|0\rangle = 0\,. \tag{5}$$

As consequence of the interaction, the bound state of H_0 vanishes, and it is replaced by a resonance that, under some conditions on the form factor $f(\omega)$, depends analytically on the coupling constant λ. This resonance is a pole of the analytic continuation through the real axis of the function

$$g(z) = \langle 1|\frac{1}{z - H}|1\rangle\,. \tag{6}$$

The function $g(z)$ is often called the reduced resolvent of H at the complex number z, not in the spectrum of H. This function is analytic with no poles on (an open set of) the complex plane with a branch cut on the positive semi-axis, admitting an analytic continuation through this cut. It is interesting to use the explicit form of its inverse $\eta(z)$ ($g(z) = 1/\eta(z)$) given by [10, 11]

$$\eta(z) = \omega_0 - z - \int_0^\infty d\omega\, \frac{\lambda^2 f^2(\omega)}{\omega - z}\,. \tag{7}$$

The function representing the boundary values of $\eta(z)$ on the positive semiaxis from above to below $\eta(\omega + i0)$ and from below to above $\eta(\omega - i0)$ are here denoted as $\eta^+(\omega)$ and $\eta^-(\omega)$

respectively. They are complex conjugate of each other. The analytic continuation of $\eta(\omega+i0)$ has a zero located at $z_R = E_R - i\Gamma/2$, the resonance pole of $g(z)$. The real part E_R of z_R is identified with the resonance energy and the imaginary part $\Gamma/2$ with the half width. Its inverse $\tau = 2/\Gamma$ is the half life. In [10], the following creation and annihilation operators are defined:

$$A^\dagger_{IN} := \int_\gamma d\omega \, \frac{\lambda f(\omega)}{\omega - z_R} \, b^\dagger_\omega - a^\dagger \,, \tag{8}$$

$$A_{OUT} := \int_\gamma d\omega \, \frac{\lambda f(\omega)}{\omega - z_R} \, b_\omega - a \,, \tag{9}$$

$$B^\dagger_{\omega,IN} := b^\dagger_\omega + \frac{\lambda f(\omega)}{\tilde{\eta}^+(\omega)} \left\{ \int_0^\infty d\omega' \, \frac{\lambda f(\omega')}{\omega' - \omega - i0} \, b^\dagger_{\omega'} - a^\dagger \right\} , \tag{10}$$

$$B_{\omega,OUT} := b_\omega + \frac{\lambda f(\omega)}{\tilde{\eta}^+(\omega)} \left\{ \int_0^\infty d\omega' \, \frac{\lambda f(\omega')}{\omega' - \omega - i0} \, b_{\omega'} - a \right\} , \tag{11}$$

where,

$$\frac{1}{\tilde{\eta}^+(\omega)} := \frac{1}{\eta^+(\omega)} + 2\pi i A \delta(\omega - z_R) \,, \tag{12}$$

In these equations γ is the circular contour around the pole and A is the residue of $1/\eta^+(\omega)$ on the pole at z_R. Operators (8-11) satisfy the following commutation relations:

$$[A_{OUT}, A^\dagger_{IN}] = 1 \, ; \qquad \frac{\eta^+(\omega)}{\eta^-(\omega)} \, [B_{\omega,OUT}, B^\dagger_{\omega',IN}] = \delta(\omega - \omega') \,. \tag{13}$$

All other commutators vanish.

The Hamiltonian H can be written in terms of these operators as:

$$H = z_R \, A^\dagger_{IN} \, A_{OUT} + \int_0^\infty d\omega \, \omega \, \frac{\eta^+(\omega)}{\eta^-(\omega)} \, B^\dagger_{\omega,IN} \, B_{\omega,OUT} \,. \tag{14}$$

the operators $A^\dagger_{IN}$ and A_{OUT} are, respectively, the creation and annihilation operators of the decaying) Gamow vector $|\psi^D\rangle$:

$$|\psi^D\rangle = A^\dagger_{IN}|0\rangle \,, \qquad A_{OUT}|\psi^D\rangle = |0\rangle \,. \tag{15}$$

From the point of view of the time asymmetric quantum theory [12, 13], both operators are defined for $t > 0$, that is the "decaying part" of a resonance process. In (15), we have seen that they play the role of creation and annihilation operators of the decaying Gamow vector, which is defined for $t > 0$ only [3].

3. The entropy for the harmonic oscillator from the path integrals approach.

The procedure to obtain the canonical entropy corresponding to a system with Hamiltonian H is well known. Consider the partition function Z for a system with time independent Hamiltonian H and let $\beta := 1/(kT)$, where k is the Boltzmann constant and T is the absolute temperature. Then, the canonical entropy is given by

$$S = k \left(1 - \beta \frac{\partial}{\partial \beta} \right) \log Z \,. \tag{16}$$

Now, let us assume that H represents the Hamiltonian of the harmonic oscillator in one dimension. Let us denote by $|n\rangle$ the eigenvectors of H, thus $H|n\rangle = \hbar\omega(n+1/2)|n\rangle$. As is well known, the partition function is now defined as

$$Z = \mathrm{tr}\,\{e^{-\beta H}\} = \sum_{n=0}^{\infty} \langle n|e^{-\beta H}|n\rangle\,. \tag{17}$$

Then, the value of the canonical entropy for the one dimensional harmonic oscillator can be directly obtained from (16). A straightforward calculation gives:

$$S = -k\log[2\sinh(\beta\hbar\omega/2)] + k\,\frac{\beta\hbar\omega}{2}\,\coth\left(\frac{\beta\hbar\omega}{2}\right)\,. \tag{18}$$

The use of path integrals to give an approximate expression for the entropy was introduced by Feynman and Hibbs [14]. Here, the idea is firstly to express the canonical ensemble in terms of coherent states , and then use the path integrals to compute the density operator. As it is well known, coherent states are defined from a vacuum state $|0\rangle$ as

$$|\alpha\rangle := e^{\alpha a^\dagger - \alpha^* a}|0\rangle\,, \tag{19}$$

where $a^\dagger$ and a are the creation and annihilation operators for the harmonic oscillator, respectively, acting on the vacuum state $|0\rangle$, which is the ground state of the harmonic oscillator.

Take now the density operator $\rho = e^{-\beta H}$ and use the strategy of path integrals to estimate its matrix elements with respect to the coherent states. This is for any pair of complex numbers α_i and α_f:

$$\langle\alpha_i|\rho|\alpha_f\rangle = \lim_{N\mapsto\infty} \rho_N(\alpha_i,\alpha_f)\,, \tag{20}$$

where

$$\rho_N(\alpha_i,\alpha_f) = \int \prod_{k=1}^{N}\left(\frac{d^2\alpha_k}{\pi}\right)\,\exp\left\{-\tau\left[\sum_{n=1}^{N} H_+(\alpha_{n-1},\alpha_n)\right.\right.$$
$$\left.\left.+\sum_{n=1}^{N+1}\left\{\left(\frac{\alpha_n^*-\alpha_{n-1}^*}{2\tau}\right)\,\alpha_n - \alpha_{n-1}^*\,\left(\frac{\alpha_n-\alpha_{n-1}}{2\tau}\right)\right\}\right]\right\}\,, \tag{21}$$

with $\alpha_0 = \alpha_i$, $\alpha_{N+1} = \alpha_f$ and $\tau = \beta/N$. We write $\alpha_i = x_i + iy_i$ and $d\alpha_i = dx_i dy_i$, so that we have $2N$ integrals in the variables $x_1,\ldots,x_N$, $y_1,\ldots,y_N$. The integration limits are $-\infty$ and ∞ in all cases, since there must be one coherent state for any complex number. The term

$$H_+(\alpha,\alpha') = \frac{\langle\alpha|H|\alpha'\rangle}{\langle\alpha|\alpha'\rangle}\,, \qquad \langle\alpha|\alpha'\rangle = \exp\left\{-\frac{|\alpha|^2}{2} - \frac{|\alpha'|^2}{2} + \alpha^*\alpha'\right\} \tag{22}$$

is called the normal expansion of the Hamiltonian H. The final result is

$$\rho(\alpha_i,\alpha_f) := \frac{1}{\langle\alpha_i|\alpha_f\rangle}\,\rho_N(\alpha_i,\alpha_f) = \exp\left\{-\frac{1}{2}\,\beta\hbar\omega\right\}\times\exp\{-\beta\hbar\omega\alpha_i^*\alpha_f\}\,, \tag{23}$$

which does not depend on N. Thus, (23) represents approximate matrix elements, in terms of the coherent states, of the canonical partition function of the harmonic oscillator.

Let us use now (16) in order to obtain an approximate expression for the entropy. First of all, let us calculate an (approximate) partition function by taking the trace of ρ as obtained in (23) with the aid of coherent states. It gives ($\alpha = x + iy$):

$$Z = \int \frac{d^2\alpha}{\pi}\, \rho(\alpha,\alpha) = \frac{1}{\pi}\, e^{-(\beta\hbar\omega)/2} \int_{-\infty}^{\infty} dx \int_{-\infty}^{\infty} dy\, e^{-\beta\hbar\omega(x^2+y^2)} = e^{-(\beta\hbar\omega)/2}\, \frac{1}{\beta\hbar\omega}\,.$$

$$\log Z = -\frac{1}{2}\,\beta\hbar\omega - \log(\beta\hbar\omega)\,, \tag{24}$$

which leads to

$$S \approx k(1 - \log(\beta\hbar\omega))\,, \tag{25}$$

which certainly gives an approximation for (18). To see it, let us note that $\coth x = 1/x + \ldots$ and $\sinh x = x + \ldots$. Applying these approximations in (18) leads to (25).

The moral of the present Section is that the use of coherent states to calculate the matrix representing the density operator is a technique amenable for extensions to systems where the standard notion of probabilities cannot be applied.

4. A notion of entropy for quantum decaying states

Now we face the problem of giving an approximate expression for the canonical entropy of a quantum system with resonances. This partition function will be find again with the use of the path integral method and coherent states, as we did for the harmonic oscillator in the previous section. The most natural basis to compute the trace is by taking the object $\{|\psi^D\rangle, |\omega^+\rangle\}$, where $|\psi^D\rangle$ is the decaying Gamow vector and $|\omega^+\rangle$ is a complete set of generalized eigenvectors of the total Hamiltonian H. The expression of these vectors in the basis $\{|1\rangle, |\omega\rangle\}$ of eigenvectors of the unperturbed Hamiltonian H_0 ($H_0|1\rangle = \omega_0|1\rangle$, $H_0|\omega\rangle = \omega|\omega\rangle$, $\omega_0 > 0$, $\omega \in [0,\infty)$ is

$$|\psi^D\rangle = |1\rangle + \int_0^\infty \frac{\lambda f(\omega)}{z_R - \omega + i0}\, |\omega\rangle\, d\omega \tag{26}$$

$$|\omega^+\rangle = |\omega\rangle + \frac{\lambda f(\omega)}{\eta^+(\omega)} \left(|1\rangle + \int_0^\infty d\omega' \frac{\lambda f(\omega')}{\omega - \omega' + i0}\, |\omega'\rangle \right)\,, \tag{27}$$

where λ and $f(\omega)$ are the coupling constant and the form factor, respectively. From (26) and (27), one sees that expressions like $\langle\psi^D|\psi^D\rangle$ or $\langle\omega^+|\omega^+\rangle$ are ill defined. In fact, since discrete and continuous subspaces of a self adjoint operator are mutually orthogonal, we have that

$$\langle 1|1\rangle = 1\,; \quad \langle 1|\omega\rangle = \langle\omega|1\rangle = 0\,; \quad \langle\omega|\omega'\rangle = \delta(\omega - \omega')\,. \tag{28}$$

Take now the expression for $|\psi^D\rangle$ in (26) and use the products in (28) to obtain

$$\langle\psi^D|\psi^D\rangle = 1 + \int_0^\infty d\omega\, \frac{\lambda\, f(\omega)}{(z_R^* - \omega - i0)(z_R - \omega + i0)}\,. \tag{29}$$

The integral represents the action of a distribution on the function $f(\omega)$. This distribution is the product of $(z_R^* - \omega - i0)^{-1}$ and $(z_R - \omega + i0)^{-1}$, which is not well defined. A similar problem would arise if we consider $\langle\omega'^+|\omega^+\rangle$.

Since in the Friedrichs model, $H|\psi^D\rangle = z_R|\psi^D\rangle$ and $H|\omega^+\rangle = \omega|\omega^+\rangle$, we conclude that a trace of the form

$$\mathrm{tr}\, e^{-\beta H} = \langle\psi^D|e^{-\beta H}|\psi^D\rangle + \int_0^\infty \langle\omega^+|e^{-\beta H}|\omega^+\rangle\, d\omega \tag{30}$$

Journal of Physics: Conference Series **578** (2015) 012006 doi:10.1088/1742-6596/578/1/012006

is not well defined. Then, we shall follow a similar approximation that the one discussed in Section III, not only to obtain an approximate value of the entropy for a system with a decaying state, but also to extent on the concept itself. As in the case of the harmonic oscillator, we define the coherent state $|\alpha\rangle$ and its bra $\langle\alpha|$, for all complex number α, as:

$$|\alpha\rangle := \exp\{\alpha\, A_{IN}^\dagger - \alpha^*\, A_{OUT}\}\,|0\rangle\,,$$
$$\langle\alpha| := \langle 0|\exp\{\alpha^*\, A_{OUT} - \alpha\, A_{IN}^\dagger\}\,, \tag{31}$$

where $|0\rangle$ is the vacuum state. Making use of the commutation relations (31) it becomes evident that these coherent states satisfy the same properties than the coherent states (19). In particular,

$$A_{OUT}|\alpha\rangle = \alpha|\alpha\rangle\;; \qquad \langle\alpha|A_{IN}^\dagger = \alpha^*\langle\alpha|\;;$$
$$\int_C \frac{d^2\alpha}{\pi}\,|\alpha\rangle\langle\alpha| = 1\;; \qquad d^2\alpha = (d\mathrm{Real}\,\alpha)(d\mathrm{Im}\,\alpha)\,, \tag{32}$$

where C denotes the field of complex numbers. The normal expansion (23) is now

$$H_+(\alpha,\alpha') = z_R\,\alpha^*\alpha'\,. \tag{33}$$

Then, instead of (23), we have the following equation

$$\rho(\alpha_i,\alpha_f) = \exp\{-\beta\, z_R\,\alpha_i^*\alpha_f\}\,, \tag{34}$$

which gives

$$Z = \frac{1}{\pi}\int_{-\infty}^{\infty} dx\, e^{-\beta\, z_R\, x^2}\int_{-\infty}^{\infty} dy\, e^{-\beta\, z_R\, y^2} = \frac{1}{\beta\, z_R}\,. \tag{35}$$

where $z_R = E_R - i\frac{\Gamma}{2}$. Finally, using (16), we arrive at the result:

$$S = k(1-\log(\beta\, z_R)) = k\left[1 - \ln(\beta\sqrt{E_R^2 + \frac{\Gamma^2}{4}}) - i\arctan\left(\frac{\Gamma}{2E_R}\right)\right]\,, \tag{36}$$

In (36) we have taken the principal branch of $\log z$. The result (36) reminds the case of the harmonic oscillator, (25), except for the presence of an imaginary term. If $\Gamma \to 0$ both results do indeed coincide, after replacing $\hbar\omega$ by E_R. The presence of a complex entropy, for the case of Gamow vectors, requires of some interpretation on the meaning of its imaginary part. The situation is quite similar to the existence of complex energy for decaying states, where the imaginary part is interpreted as the inverse of the half life. Note that the resonance in the Friedrichs model is caused by the interaction of the system with the background, which plays the role of the thermodynamical bath. Then, we suggest that the real part of the entropy (36) is the entropy of the system and that the imaginary part of it is the entropy transferred from the system to the background. Should the thermodynamical entropy be identified with the modulus of (36), one concludes that the total entropy for a decaying state is bigger than the entropy of a stable system.

Journal of Physics: Conference Series **578** (2015) 012006 doi:10.1088/1742-6596/578/1/012006

5. Conclusions.

In this work we have pointed out to some of the difficulties concerning the application of concepts of Statistical Mechanics to complex-energy vectors. We have presented a suitable alternative to the probabilistic description, by implementing a representation of the decaying vectors, obtained in the framework of the Friedrisch model, written them in terms of coherent states and by performing a path integration over these states to get the density matrix operator. The results are quite encouraging, because at the level of approximation used to calculate the density operator, we do not have to introduce some ad-hoc notions like complex temperatures or treat independently real and imaginary entropies. We think that this is a promising first step towards a novel formulation of the statistical mechanics for decaying systems. We hope to be able to complete this program in the near future.

Acknowledgements.

This work has been partially supported by the CONICET of Argentina (PIP 0740) and the Spanish Government Grants No. PR2011-0343, No. FPA2008- 04772-E, and No. MTM2009-10751.

References

[1] Bohm A 2001 *Quantum Mechanics: Foundations and Applications* (Berlin and New York: Springer)
[2] Gelfand I M and Vilenkin N 1964 *Generalized Functions* (Vol. 4) (New York: Academic)
[3] Civitarese O and Gadella M 2004 *Phys. Rep* **396** 41
[4] Kobayashi T and Shimbori T 2001 *Phys. Rev.* E **63** 056101
[5] Kobayashi T 2003 *Int. J. Theor. Phys.* **42** 2265
[6] Misra B, Prigogine I and Courbage M 1979 *Proc. Natl. Acad. Sci. USA* **76** 4768
[7] Friedrichs K O 1948 On the perturbation of continuous spectra *Commun. Pure Appl. Math.* **1** 361
[8] Civitarese O and Gadella M 2013 *Found. Phys.* **43** 1275
[9] Civitarese O and Gadella M 2014 Physica A in press
[10] Antoniou I E, Gadella M, Karpov E, Prigogine I and Pronko G 2001 *Chaos, Solitons & Fractals* **12** 2757
[11] Antoniou I and Prigogine I 1993 *Physica* A **192** 443
[12] Bohm A, Kielanowski P and Wickramasekara S 2006 *Annals of Physics* **321** 2299
[13] Bohm A, Gadella M and Kielanowski P 2011 *SIGMA* **7** 086
[14] Feynman R P and Hibbs A R 1965 *Quantum Mechanics and Path Integrals* (New York: McGraw-Hill)

Status and perspectives of double beta decay searches

Kai Zuber
Institut für Kern- und Teilchenphysik, TU Dresden, 01069 Dresden, Germany
E-mail: zuber@physik.tu-dresden.de

Abstract. Double beta decay is an extremely rare process and requires half-live measurements around 10^{20} years for the neutrino accompanied mode, while for the neutrino-less mode much longer half-lives have to be explored. The various experimental approaches, currently considered for the search of this process, results will be presented.

1. Introduction and physics

Neutrinos play a crucial role in modern particle, nuclear and astrophysics including cosmology. It has been the major achievement of the last 15 years to show that neutrinos have a non-vanishing rest mass. The evidence arises from a deficit of upward going atmospheric muon neutrinos confirmed by long baseline accelerator experiments and the solution of the solar neutrino problem being confirmed by nuclear reactor measurements. All observations can be explained by neutrino oscillations, which are depending on mass differences $\Delta m_{ij}^2 = m_j^2 - m_i^2$ with $i, j = 1, 2, 3$. The determination of absolute neutrino masses is now a major issue, because neutrino oscillation experiments do not allow this. The classical way to search for a rest mass of the neutrino is the study of the endpoint region of electron spectra in beta decay (see [1] for a recent review). The KATRIN experiment is well on its way to improve the current bound of about 2.2 eV for $\bar{\nu}_e$ by an order of magnitude. Further bounds on the total sum of neutrino masses can be obtained from cosmological studies, dominated by recent Planck measurements combined with large scale structure information based on BAO suggesting that the mass is less than about 0.25 eV [2]. Another laboratory process to search for neutrino masses is the rare nuclear decay of neutrino-less double beta decay

$$(Z, A) \rightarrow (Z+2, A) + 2e^- \quad (0\nu\beta\beta\text{-decay }) \tag{1}$$

accompanied by the Standard Model process of

$$(Z, A) \rightarrow (Z+2, A) + 2e^- + 2\bar{\nu}_e \quad (2\nu\beta\beta\text{-decay }). \tag{2}$$

Single beta decay must be forbidden or at least strongly suppressed to observe this decay and only 35 potential double beta emitter exist in nature. As can be seen from Equation 1 the given decay mode is violating total lepton number by two units and thus is not allowed in the Standard Model. Being a decay the observable is a half-life which can be linked to the quantity of interest ϵ via

$$(T_{1/2}^{0\nu})^{-1} = G_{PS} \mid M_{Nuc} \mid^2 \epsilon^2 \tag{3}$$

Journal of Physics: Conference Series **578** (2015) 012007 doi:10.1088/1742-6596/578/1/012007

with G_{PS} being the phase space, $\mid M_{Nuc} \mid$ the involved nuclear matrix element for the physics process considered to describe this decay and ϵ the quantity of interest. Various beyond the Standard Model processes can be considered like light and heavy Majorana neutrino exchange, right-handed weak currents, R-parity violating SUSY (λ'_{111}) and double charged Higgs bosons. Independent of the dominant decay mechanism, if $0\nu\beta\beta$-decay is ever observed experimentally it will imply that neutrinos are Majorana particles [3]. Furthermore, given all the possible processes it will become an important question what the individual contributions of the considered processes will be. The LHC will help to restrict this by performing searches for new particles in the TeV range. An example how this complementary information from LHC and $0\nu\beta\beta$-decay can be used within the context of left-right symmetric theories is given in [4]. For a recent extensive review on the particle physics in double decay see [5].
The standard interpretation considered here is the one using light Majorana neutrino exchange. In this case ϵ is the effective Majorana neutrino mass $\langle m_{ee} \rangle$ given by

$$\epsilon \equiv \langle m_{ee} \rangle \ = \mid \sum_i U_{ei}^2 m_i \mid \tag{4}$$

with U_{ei}^2 as the PMNS-mixing matrix elements containing the electron neutrino.

2. Double beta decay and neutrino oscillation results

For the following discussion a restriction to the light Majorana neutrino case is done. It is evident from Eq. 4 that the expectation for $\langle m_{ee} \rangle$ in double beta decay depends on the neutrino oscillation parameters. It should be noted that the mixing matrix of relevance is given by

$$U = U_{PMNS}(\theta_{12}, \theta_{13}, \theta_{23}, e^{i\delta}) \times diag(1, e^{i\alpha}, e^{i\beta}) \tag{5}$$

with the standard leptonic mixing matrix U_{PMNS} and two additional CP-phases α and β, called Majorana phases, which appear if a neutrino is its own antiparticle. These phases do not show up in oscillation experiments. Hence, $\langle m_{ee} \rangle$ can be written as a sum of three terms

$$\langle m_{ee} \rangle \ = \mid m_e^1 \mid + \mid m_e^2 \mid e^{2i\alpha} + \mid m_e^3 \mid e^{2i\beta} \tag{6}$$

with the individual contributions given as

$$\begin{aligned} m_e^1 &= \mid U_{e1} \mid^2 m_1 = m_1 \cos^2\theta_{12} \cos^2\theta_{13} \\ m_e^2 &= \mid U_{e2} \mid^2 m_2 = m_2 \sin^2\theta_{12} \cos^2\theta_{13} \\ m_e^3 &= \mid U_{e3} \mid^2 m_3 = m_3 \sin^2\theta_{13} \end{aligned} \tag{7}$$

Latest global fits to available oscillation parameters are given in [6, 7]. An important new ingredient are 3-flavour fits to solar neutrinos [8], observations from T2K [9] and MINOS [10] as well as first direct measurements of θ_{13} from the reactor experiments Double Chooz, Daya Bay and RENO [11, 12, 13, 14]. As the oscillations do not fix the absolute scale there are two options for arranging the mass eigenstates, either $m_3 > m_2 > m_1$ (normal hierarchy, NH) or $m_2 > m_1 > m_3$ (inverse hierarchy,IH). If the neutrino masses turn out to be close to the current limit from beta decay there is a quasidegeneracy ($m_1 \approx m_2 \approx m_3 \equiv m_0$). In case of hierarchies the two larger masses can be expressed as

$$\begin{aligned} m_2 &= \sqrt{m_1^2 + \Delta m_\odot^2} \qquad m_3 = \sqrt{m_1^2 + \Delta m_{atm}^2} \qquad \text{(normal)} \\ m_2 &= \sqrt{m_3^2 + \Delta m_\odot^2 + \Delta m_{atm}^2} \qquad m_1 = \sqrt{m_3^2 + \Delta m_{atm}^2} \qquad \text{(inverted)} \end{aligned} \tag{8}$$

with the solar splitting $\Delta m_\odot^2 = 7.59^{+0.20}_{-0.18} \times 10^{-5} eV^2$ and the atmospheric splitting $\Delta m_{atm}^2 = 2.49 \pm 0.09 \times 10^{-3} eV^2$ (normal) , $\Delta m_{atm}^2 = -2.343^{+0.10}_{-0.09} \times 10^{-3} eV^2$ (inverted) [6]. The behaviour

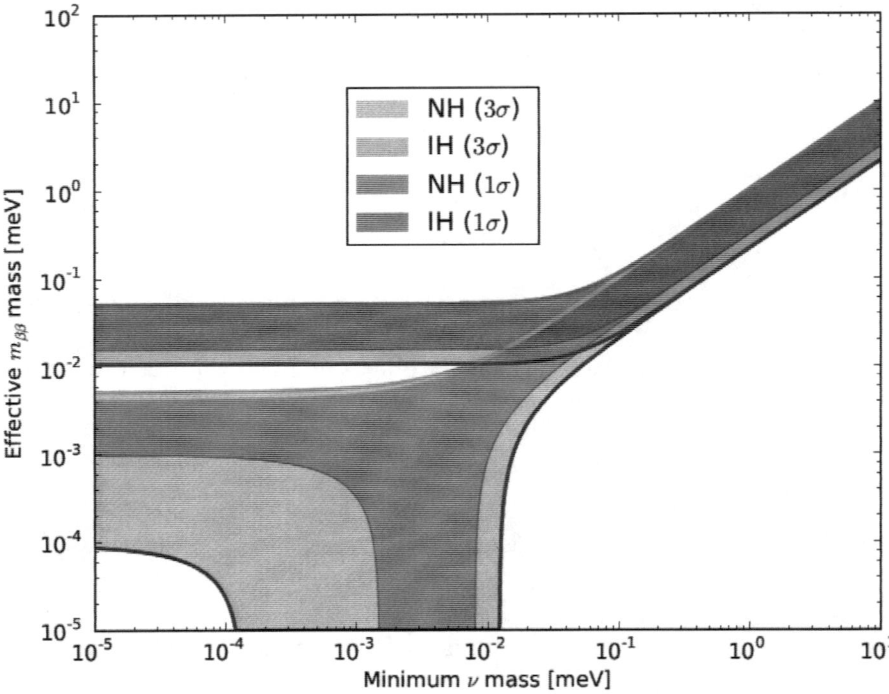

Figure 1. A presentation (as introduced by [15]) of the effective Majorana neutrino mass $\langle m_{ee} \rangle$ as a function of the lightest mass eigenstate. Shown are the inverted (green) and normal (red) hierarchy bands using the current values of the mixing angles and their 1σ and 3σ range. As can be seen within the normal hierarchy there can be perfect cancellation of terms in case that the lightest neutrino mass is in the range $10^{-3} - 10^{-2}$ eV (1σ) or even $10^{-4} - 10^{-2}$ eV (3σ).

of $\langle m_{ee} \rangle$ as a function of the lightest mass eigenstate using the current oscillation results is shown in Figure 1.

For $\langle m_{ee} \rangle$ values larger than about 100 meV neutrinos are almost degenerate, the inverted hierarchy covers a range between about 10-50 meV and below 10 meV is the region of the normal hierarchy. As can be seen, in the NH there is a chance for cancellation among the terms (Fig. 1). There is no such effect in the IH because of the non-maximal solar mixing angle θ_{12}. Half-lives for the IH are in the region beyond 10^{26}years while half-lives in the NH are well beyond 10^{28} years.

3. General experimental considerations

Historically three different kinds of methods have been used, radiochemical (basically searching for the double beta decay of ^{238}U), geochemical (exploring billion years old ores) and direct searches. Nowadays the first two methods are of less importance as they cannot distinguish between the different decay modes and are considered to be dominated by $2\nu\beta\beta$-decay , Equation (2).

Evidently measurements of half-lives around 10^{20} years and well beyond are by no means trivial. As signal for the process given in Equation (1) serves a peak in the sum energy spectrum of the two electrons equivalent to the Q-value of the nuclear transition. From Equation (1) it is also apparent that the maximal information obtainable from measurements will be the single electron energies, the opening angle between them and the daughter ion. This might be accompanied by

Journal of Physics: Conference Series **578** (2015) 012007 doi:10.1088/1742-6596/578/1/012007

characteristic gamma rays in case of excited state transitions. The corresponding half-life in case of no background is given by the radioactive decay law (assuming a measuring time $t \ll T^{0\nu}_{1/2}$)

$$T^{0\nu}_{1/2} = \ln 2 matN_A/N_{\beta\beta} \tag{9}$$

with m the used mass, a the isotopic abundance of the double beta emitter, t the measuring time, N_A the Avogadro constant and $N_{\beta\beta}$ the number of double beta events, which has to be taken from the experiment. If no peak is observed and a constant background (in general terms these are all potential energy depositions in the region of interest, i.e. around the Q-value, not being neutrino-less double beta decay) is assumed scaling linearly with time, a half-live estimate can be derived as

$$(T^{0\nu}_{1/2})^{-1} \propto a \times \epsilon \sqrt{\frac{M \times t}{B \times \Delta E}} \tag{10}$$

where ϵ is the efficiency for detection of the total energy of both electrons, ΔE is the energy resolution at the peak position and B the background index normally given in counts/keV/kg/year. Hence, the most crucial parameters are a high detection efficiency and high abundance of the isotope of interest. This is the reason why almost all next generation experiments are using enriched materials and the "source = detector" approach, i.e. the emitter is part of the detector itself. Furthermore, the energy resolution[1] should be as good as possible to concentrate the few expected events in a small region and ideally the experiment should be background free. An irreducible background is the Standard Model process $2\nu\beta\beta$-decay . Here

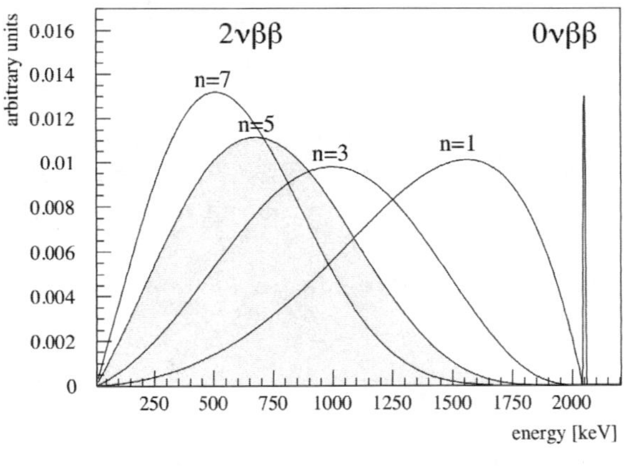

Figure 2. Schematic plot of the sum energy spectrum of the two electrons in double beta decay, shown here for ^{76}Ge . $0\nu\beta\beta$-decay results in a peak at the Q-value of the transition. Various modes can be characterised by the phase space dependence (Q-E)^n. The mode n=5 is the $2\nu\beta\beta$-decay while the modes n=1,3,7 involve the emission of a majoron, a Goldstone boson linked to the spontaneous breaking of lepton number. The different modes belong to different behaviours of the majoron with respect to weak isospin. The individual contributions are not to scale.

again energy resolution matters, because of the continuous spectrum of the $2\nu\beta\beta$-decay mode, its high energy part is leaking into the peak region (see Figure (2)). Nevertheless, this can be

[1] Care must be taken when comparing experiments, as for traditional reasons different detector technologies use either the Gaussian σ or the Full Width at Half Maximum ΔE to quote energy resolution. The relation among the quantities is $\Delta E = 2.35\sigma$.

Journal of Physics: Conference Series **578** (2015) 012007 doi:10.1088/1742-6596/578/1/012007

a worry as the half-life is typically several orders of magnitude shorter than the expected one for $0\nu\beta\beta$-decay . As the decay rate for $0\nu\beta\beta$-decay scales with Q^5 only isotopes with Q-values above 2 MeV are considered for experimental searches. They are listed together with their natural abundance and Q-value in Table (1).

Isotope	nat. abund. (%)	Q-value (keV)	Experiment
^{48}Ca	0.187	4267.96 ± 0.32	CANDLES
^{76}Ge	7.8	2039.006 ± 0.050	GERDA, MAJORANA
^{82}Se	9.2	2995.5 ± 1. 9	SuperNEMO, LUCIFER
^{96}Zr	2.8	3347.7 ± 2.2	-
^{100}Mo	9.6	3034.40 ± 0.17	AMoRE
^{110}Pd	11.8	2017.85 ± 0.64	-
^{116}Cd	7.5	2813.50 ± 0.13	COBRA, $CdWO_4$
^{124}Sn	5.64	2292.64 ± 0.39	(Tin.Tin)
^{130}Te	34.5	2527.518 ± 0.013	CUORE, SNO+
^{136}Xe	8.9	2457.83 ± 0.37	EXO, KamLAND-Zen, NEXT
^{150}Nd	5.6	3371.38 ± 0.20	MCT

Table 1. Table showing the eleven candidate isotopes with a Q-value larger than 2 MeV. Given are the natural abundances and Q-values as determined from precise Penning trap measurements or from the Atomic Mass Evaluation 2012 [16]. The last column shows the experiments addressing the measurement of the corresponding isotope. For some experiments only the "default" isotope is mentioned as they consider exploring several ones. Additionally, several other research and development projects are ongoing.

As mentioned, most experiments follow the approach that the source is equal to the detector, i.e. building a detector which contains the isotope of interest. Technologies used for that are semiconductors, cryogenic bolometers, scintillators and liquid noble gas detectors. The alternative is to use tracking devices in form of TPCs containing thin foils of double beta emitters. Here single electron spectra and opening angles can be measured as well.

4. Experimental status

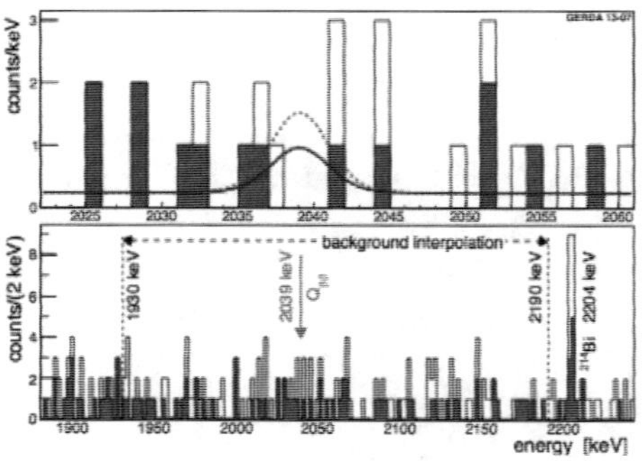

Figure 3. Measured spectrum of the GERDA experiment in phase I. The search was performed as a blind analysis. No apparent peak is visible and the result strongly disfavors a claim made 10 years ago (from [19]).

The last years were dominated to set-up experiments being able to probe the claimed observation of $0\nu\beta\beta$-decay in ^{76}Ge [17]. The start has happened with new double beta experiments, namely GERDA (using ^{76}Ge), EXO and KamLAND-Zen (using ^{136}Xe) and CANDLES (using ^{48}Ca). GERDA [18] is a next generation experiment based on Ge-semiconductors consisting of Ge-crystals enriched in ^{76}Ge . The experiment is located in the Gran Sasso Underground Laboratory (Italy). The idea is to run the bare crystals within LAr which serves as shielding and cooling. In a first phase a total of eight isotopically enriched detectors (from the former Heidelberg-Moscow and IGEX experiments) with a total mass of 17.7 kg and three natural Ge-detectors have been deployed and official data taking started on 1. Nov. 2011. The spectrum in the region of interest at the end of phase 1 is shown in Figure (3). Apparently the data do not support the claim [19]. Currently the experiment is preparing for phase 2, which is aiming to reduce background by another order of magnitude and adding more detectors.
Due to the relative cheap enrichment of noble gases, two large scale experiments based on enriched ^{136}Xe have started. First is EXO-200, using about 175 kg LXe in form of a TPC located at WIPP (USA). Also EXO-200 could not find a peak and provide a lower limit of 1.1×10^{25} yrs (90 % CL) [20]. A second Xe approach is KamLAND-Zen using the well understood infrastructure of the KamLAND experiment. To perform double beta decay with Xe-loaded liquid scintillator a special mini-balloon was constructed and deployed within KamLAND. In this way about 330 kg of enriched Xe could be filled in the detector and data taking has started in September 2011. A lower half-life limit for the neutrinoless decay of 1.9×10^{25} yrs (90 % CL) is given [21]. Last but not least there is CANDLES using 305 kg of CaF_2 scintillators, focusing on ^{48}Ca , the isotope with the highest Q-value of all double beta emitters. The experiment is installed in the Kamioka mine (Japan) and data taking has started recently.
As the evidence is not confirmed, the next goal will be to reach the region of the inverted hierarchy, i.e $\langle m_{ee}\rangle$ below $\approx$ 50 meV.The requirement on the necessary half-life to touch this region is given in [22]. As the requirements on mass and background for this purpose are already very demanding, fully excluding the IH is more than challenging. If there is no signal found in the IH the ultimate step is the exploration of the normal hierarchy. However, for this ton scale experiments with extraordinary low background have to be considered and completely new background components have to taken into account, for example neutrino-electron scattering due to solar ^{8}B neutrinos [23]. For that half-lives well beyond 10^{28} yrs have to be measured.

5. Alternative processes including positrons and electron capture

An equivalent process to the one discussed is $\beta^+\beta^+$-decay also in combination with electron capture (EC). There are three different variants possible depending on the Q-value:

$$(Z,A) \rightarrow (Z-2,A) + 2e^+(+2\nu_e) \quad (\beta^+\beta^+) \tag{11}$$
$$e_B^- + (Z,A) \rightarrow (Z-2,A) + e^+(+2\nu_e) \quad (\beta^+/EC) \tag{12}$$
$$2e_B^- + (Z,A) \rightarrow (Z-2,A)(+2\nu_e) \quad (EC/EC) \tag{13}$$

$\beta^+\beta^+$ is always accompanied by EC/EC or β^+/EC-decay. The positron production reduces the effective Q-value by $2m_ec^2$ per positron. Therefore, the rate for $\beta^+\beta^+$ is small and energetically only possible for six nuclides, however it would have a striking signature with four 511 keV gamma rays. It was shown that the β^+/EC-mode has an enhanced sensitivity to right-handed weak currents [24] and might be valuable to explore if $0\nu\beta\beta$-decay is discovered. The full Q-value is available in the EC/EC mode which is the hardest to detect experimentally. However, it was proposed [25, 26] that if an excited state of the daughter nucleus is degenerate with the original ground state a resonance enhancement in the decay rate could occur and the de-excitation gammas would serve as a nice signal. Due to the sharpness of the resonance a more detailed study of candidates had to wait for Penning traps entering the field and exploring reasonable candidates. The most reliable one seems to be ^{152}Gd (see Figure 4) where such a scenario is realised [27]. Despite this nice effect, to achieve the same sensitivity of $\langle m_{ee} \rangle$ as in $0\nu\beta\beta$-decay seems to require a measurement of the half-life an order of magnitude longer making this method slightly less attractive. Decays have been searched for in a number of isotopes with half-life limits ranging from $10^{15} - 10^{21}$ years, but no signal could be found in direct searches. However there is a weak evidence from geochemical experiments for a an effect in the ^{130}Ba system [28, 29].

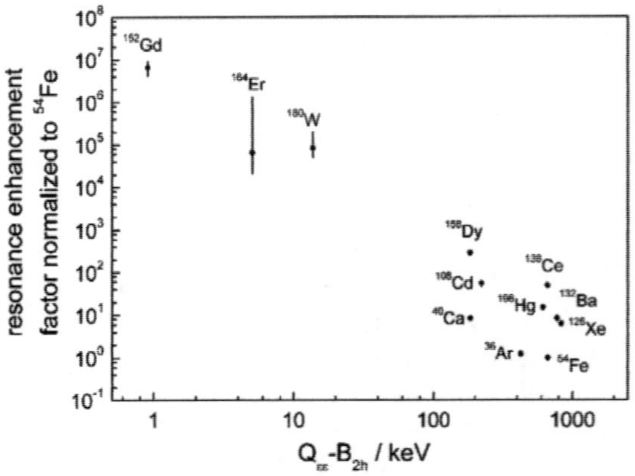

Figure 4. Compilation of enhancement factors in resonant neutrino-less double electron capture based on mass measurements with Penning traps (from [27]).

References

[1] Otten E and Weinheimer C 2008 *Rep. Prog. Phys.* **71** 086201
[2] Ade P A R *et al* 2013 *Preprint* arXiv:1303.5076, arXiv:1311.1657
[3] Schechter J and Valle J W F 1981 *Phys. Rev.* D **23** 1666
[4] Tello V *et al* 2011 *Phys. Rev. Lett.* **106** 151801
[5] Rodejohann W 2011 *Int. Journal of Modern Physics* **20** 1833
[6] Schwetz T, Tortola M and Valle J W F 2011 *New J. Phys.* **13** 063004
[7] Fogli G L *et al* 2011 *Phys. Rev.* D **84** 053007
[8] Aharmim B *et al* 2013 *Phys. Rev.* C **88** 025501
[9] Abe K *et al* 2011 *Phys. Rev. Lett.* **107** 041801
[10] Adamson P *et al Phys. Rev. Lett.* **110** 251801
[11] Abe Y *et al* 2012 *Phys. Rev. Lett* **108** 131801
[12] An F P *et al* 2012 *Phys. Rev. Lett.* **108** 171803
[13] Ahn J K *et al* 2012 *Phys. Rev. Lett.* **108** 191802
[14] Abe Y *et al* 2012 *Preprint* arXiv:1207.6632
[15] Vissani F 1999 *JHEP* **9906** 022
[16] Wang M *et al* 2012 *Chin. Phys.* C **36** 1603
[17] Klapdor-Kleingrothaus H V *et al* 2004 *Phys. Lett.* B **578** 54
[18] Ackermann K-H *et al* 2013 *Eur. Phys. J.* C **73** 2330
[19] Agostini M *et al* 2013 *Phys. Rev. Lett.* **111** 122501
[20] Albert J B *et al* 2014 *Preprint* arXiv:1402.6956
[21] Gando A *et al* 2012 *Phys. Rev.* C **85** 045504
[22] Dueck A, Rodejohann W and Zuber K 2011 *Phys. Rev.* D **83** 113010
[23] deBarros N F and Zuber K 2011 *J. Phys. G: Nucl. Phys.* **38** 105201
[24] Hirsch M *et al* 1994 *Z. Phys.* A **347** 151
[25] Bernabeu J, DeRujula A and Jarlskog C 1983 *Nucl. Phys.* B **223** 15

[26] Sujkowski Z and Wycech S 2004 *Phys. Rev.* C **70** 052501
[27] Eliseev S *et al* 2011 *Phys. Rev. Lett.* **106** 052504
[28] Barabash A and Saakyan R 1996 *Phys. Atom. Nucl.* **59** 179
[29] Meshik A P *et al* 2001 *Phys. Rev.* C **64** 035205

Journal of Physics: Conference Series **578** (2015) 012008 doi:10.1088/1742-6596/578/1/012008

Neutron stars with dark energy

P O Hess[1], I Rodríguez[2], W Greiner[2] and T Boller[3]

[1] Instituto de Ciencias Nucleares, UNAM, Circuito Exterior, C.U., A.P. 70-543, 04510 México, D.F., México

[2] Frankfurt Institute for Advanced Studies, Johann Wolfgang Goethe Universität, Ruth-Moufang-Str. 1, 60438 Frankfurt am Main, Germany

[3]Max-Planck Institute for Extraterrestrial Physics, Giessenbachstrasse 1, 85748 Garching, Germany

E-mail: hess@nucleares.unam.mx

Abstract. After a short review on the possible experimental observations to verify pseudo-complex General Relativity, neutron stars as a particular object of interest are investigated. Dark energy is added to the structure of a neutron star, while for the nuclear part the chiral SU(3) model is used. For the coupling of matter to dark energy a special assumption is made. The consequences are discussed. We show that neutron stars of up to six solar masses are obtained, which already behave similar to a black hole.

1. Introduction

Neutron stars are the results of supernova explosions leading to highly compact objects, whose densities are of the order or several times of the nuclear density. In order to understand these compact stars, the equation of state for nuclear matter has to be known. In [1] the chiral $SU(3)$ model was developed which includes the contributions of hadronic resonances, scalar and vector mesons. The model predicts for the maximal mass of a neutron star 2.05 solar masses. Larger stars should contract to a black hole.

In [2, 3] a modified theory of General Relativity, the *pseudo-complex General Relativity* (pc-GR), was presented. The novel feature is that it includes a minimal length. But even when the limit of zero minimal length is considered, an additional contribution of a dark energy remains. This dark energy is repulsive and finally avoids the formation of a black hole. The event horizon also disappears and thus no black hole exists in this theory rather very high dense objects are possible, though for a distant observer they look similar to a black hole.

It is therefore interesting to describe these highly dense objects within the new theory. For that the properties of matter at very high densities have to be known, which we do not at this moment. Nevertheless, as a starting point one can assume that the chiral $SU(3)$ model is still valid and take further assumptions on the coupling of mass to the dark energy. The important point is to see if this dark energy is repulsive enough to avoid the formation of a black hole, to investigate its properties and observational consequences and what is the highest mass possible. Of course, the results will depend on the knowledge of the properties of nuclear matter at high densities, which is not well known.

In what follows, we will first resume the main properties of the pc-GR and present some observational consequences. We show that standard GR has problems to describe *Quasi Periodic*

Objects (QPO), which form in accretion disks around galactic black holes, while pc-GR is able to explain the observations. After that, neutron stars are described and we will show that under certain assumptions neutron stars of up to six solar masses are produced. The larger the mass the more it resembles a black hole, *but never it is a black hole.*

2. Properties of pc-GR and observations

In pc-GR instead of real coordinates X^μ (μ=0,1,2,3) pseudo-complex coordinates are introduced [2, 3]. Pseudo-complex means that the coordinate is given by $X^\mu = x^\mu + Iy^\mu$ where x^μ is the standard real coordinate and y^μ the pseudo-imaginary component. The label *pseudo-complex* is due to $I^2 = 1$, which has important mathematical and physical consequences. The imaginary component is proportional to the four-velocity and due to dimensional reasons (that the unit of the pseudo-imaginary component is equal to the real component) a minimal length parameter appears as a factor, i.e. $y^\mu = \frac{l}{c}u^\mu$. This structure may have important consequences when one tries to quantize the theory, but for our purpose (large objects) one can neglect the contribution of the minimal length, i.e. it is set to zero. But there is another consequence of the extended theory: It requires the presence of a dark energy which is repulsive.

The origin of the dark energy is probably due to vacuum fluctuations, as was shown in [4]. There, a calculation within semi-classical quantum mechanics shows that mass produces quantum fluctuations in the neighborhood of a Schwarzschild black hole, which increases with its mass and falls off as $1/r^6$, with corrections. Semi-classical quantum mechanics means that the back ground metric is fixed, i.e. the event horizon remains. In [5] this result is used to speculate on the possibility that the collapse of a star is halted before the black hole is formed, when the collapse is "slow" enough. The disadvantage of this theory is that it can not determine the action of the dark energy on the metric. In pc-GR the dark energy is considered as a classical fluid and its back-reaction on the metric can be determined, with the result that the event horizon is avoided (the metric component g_{00} is always positive).

From these observations a new principle is proposed: *Mass not only curves the space around it but also changes the vacuum properties near a mass, which in turn changes the metric.*

There are already some observations that pc-GR may be a correct extension to GR, namely the QPO in galactic black holes [6, 7, 8]. These are binary systems which contain an assumed black hole with a stellar partner, which provides mass to the accretion disk. In the inner boarders of the accretion disk QPO may form, which are local eruptions. These QPO follow the rotation of the disk and emit Fe-Kα lines, from which the red-shift can be deduced. The GR and pc-GR predict the dependence of the orbital frequency and the red-shift on the distance to the center of the dense object. In a consistent theory, from both observations the same radial distance has to result. In [9] the dependence of the orbital frequency and the red-shift are determined for GR and pc-GR. In GR the orbital frequency always increases toward smaller distance while pc-GR exhibits a maximum from which on it decreases again. As a consequence, low orbital frequencies are possible for small radial distances. The red-shift, at the equator, shows in both theories a similar behavior, however in pc-GR the curve is shifted to smaller radial distances and it never reaches infinity, though it assumes very large values. The observations show [6, 7, 8] that assuming GR there is a striking difference in the deduced radial distance from the orbital frequency and the red-shift. Using the observed orbital frequency a distance between 8 to 16 Schwarzschild radii are observe for the system XTE J-1550-564, which corresponds to a compact star with 9.10±0.15 solar masses [10]. while from the observed red-shift a distance between 2-3 Schwarzschild radii is obtained! Similar for all other systems observed up to now. In pc-GR, however, the same radial distance is obtained. Assuming Keplerian Motion and that QPO's and relativistic iron lines originate from the same region, *this is the first observation which may confirm pc-GR*! It also gives us an argument to continue the study neutron stars.

3. Neutron stars within pc-GR

In order to describe neutron stars, the modified Einstein equations are used, which include on the right hand side of these equations not only the baryonic energy-momentum tensor but also the contribution due to the dark energy. In the limit of zero minimal length, the equations are

$$\mathcal{R}^{\mu i}_{\nu} - \frac{1}{2} g^{\mu i}_{\nu} \mathcal{R}^{i} = -\frac{8\pi k}{c^4} T^{\Lambda \mu i}_{\ \ \nu} - \frac{8\pi k}{c^4} T^{m \mu}_{\ \ \nu} \tag{1}$$

(i stands for the interior region of the neutron star) where $\mathcal{R}^{\mu}_{\nu}$ denotes the real projection of the Ricci tensor, $\mathcal{R}$ the real projection of the Ricci scalar and g^{μ}_{ν} the corresponding projection of the metric.

For both, the matter and the dark energy, we assume an isotropic fluid, though it can be easily extended to an anisotropic fluid. With this assumption, the Einstein equations can be rewritten in terms of the Tolman-Oppenheimer-Volkov (TOV) equations [11] (for details, please consult [12])

$$\begin{aligned}
\frac{dp_m}{dr} &= -\frac{(\varepsilon_m(r) + p_m(r))}{r[r - 2m_m(r) + 2m_\Lambda(r)]} \times \\
&\quad \left[m_m(r) - m_\Lambda(r) + \frac{4\pi k}{c^4} r^3 (p_\Lambda(r) + p_m(r)) \right] \\
\frac{dp_{\Lambda i}}{dr} &= -\frac{(\varepsilon_{\Lambda i}(r) + p_{\Lambda i}(r))}{r[r - 2m_m(r) + 2m_{\Lambda i}(r)]} \times \\
&\quad \left[m_m(r) - m_{\Lambda i}(r) + \frac{4\pi k}{c^4} r^3 (p_{\Lambda i}(r) + p_m(r)) \right]
\end{aligned} \tag{2}$$

p_m is the matter pressure and ε_m the matter density. The $p_{\Lambda i}$ is the pressure of the dark energy density in the internal region of the neutron star and $\varepsilon_{\Lambda i}$ its density. The m_m and $m_{\Lambda i}$ is the accumulated mass of the matter and dark energy as a function of radial distance. They are obtained by integrating over the distance.

These equations are not complete and the equation of state of the matter and the dark energy has to be known yet. For the matter the chiral-$SU(3)$ model [1] is used. It is a mean field model, which not only takes into account the nucleon fields (proton and neutron) but also baryon resonances (Λ, Σ and Ξ) and their coupling to several scalar and vector mesons. The model includes several parameters which where fixed to properties of nuclear matter, as for example binding energies. The resulting equation of state is shown in Fig. 1. As can be seen the dependence of the pressure versus energy density is not linear in all the region.

Assumptions have to be made on the unknown coupling of the matter density to the dark energy density. We take the simplest assumption, as done in [13], which corresponds to a linear coupling

$$\varepsilon_{\Lambda i} = \alpha \varepsilon_m \quad . \tag{3}$$

This is a crude approximation and may need modifications, as we will see later on. In difference to [13] the proportionality factor α is negative, i.e. the dark energy density is negative.

Having defined the equation of states, we proceed in solving numerically the TOV equations. A 4th order Runge-Kutta algorithm is applied, defining at the center of the star initial pressure and density values.

The integration starts at the center and going toward larger r. The initial values for the baryon density varies from one to several times the nuclear density. The one for the dark energy

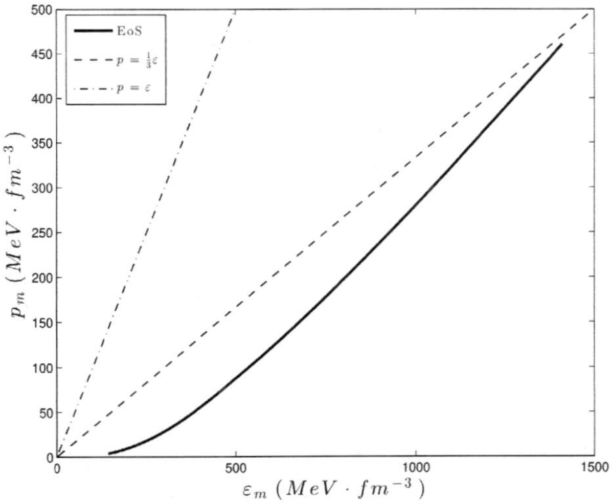

Figure 1. Equation of state, pressure versus energy density, for nuclear matter. Two further curves are shown for comparison, one for radiation and the other with $p_m = \varepsilon_m$.

is fixed by the proportionality relation assumed. The integration stops when the baryonic pressure is zero (for numerical reasons it is set to be less than 10^{-11} in units of the pressure). This defines the surface of the star! The results vary with the change of the central baryonic density and pressure, but the general structure keeps the same. of course, the central baryonic density can not be arbitrarily large. The integration is not performed from the outside to the inside, because this would require the knowledge of the position of the surface. This is the procedure as explained in more detail in the book by N. K. Glendenning [14].

One of the results is shown in Fig. 2, where the total mass of neutron stars is plotted versus their radii. The stable branch is given by the negative slope part of the curve. As is seen, stars with up to six solar masses are obtained. The larger the absolute value of the proportionality factor α (stronger repulsion) the larger the masses possible for neutron stars.

In Fig. 3 the baryonic density is plotted versus the radial distance. The larger the absolute value of α the further the density distribution reaches outside, i.e. the mass has to increase because the surface under the curve increases. The radius of the star is obtained when the *pressure* lowers below a fixed but very small value.

As a last result, we plot the so-called *compactness* in Fig. 4. The compactness is defined as $\frac{2M_m}{R}$, where M_m is the total baryonic mass and R the radius of the star. The definition originates in standard GR where the g_{00} component of the Schwarzschild solution is $\left(1 - \frac{2M_m}{r}\right)$. When $r = R$ reaches the Schwarzschild radius, thus values is zero, i.e. a value near to 1 indicates the onset of the forming of a black hole. In pc-GR it gives an idea on how near the star resembles a black hole. If the value is near to 1, the red-shift acquires large values and for a distant observer the object resembles a black hole, *though it is not a black hole.*

Fig. 4 demonstrates that the neutron stars near six solar masses resemble more black holes than normal neutron stars.

Larger masses of neutron stars can not be obtained with the assumption of a linear coupling between baryonic and dark energy density. When $\mid \alpha \mid$ is greater than 1, the fall off of the dark energy density toward the radius of the star is not quick enough and the repulsion is so large that it starts to evaporate the surface, avoiding the accumulation of more mass. In order to resolve this problem, one has to use a different coupling between the baryonic and dark energy

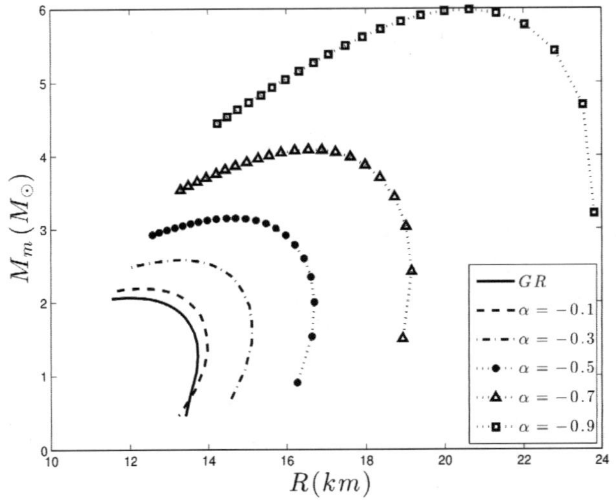

Figure 2. Baryonic total mass vs. total radius for different values of the coefficient α. The central Λ-term pressure $p_{\Lambda c}$ has been fixed to $1\varepsilon_0$.

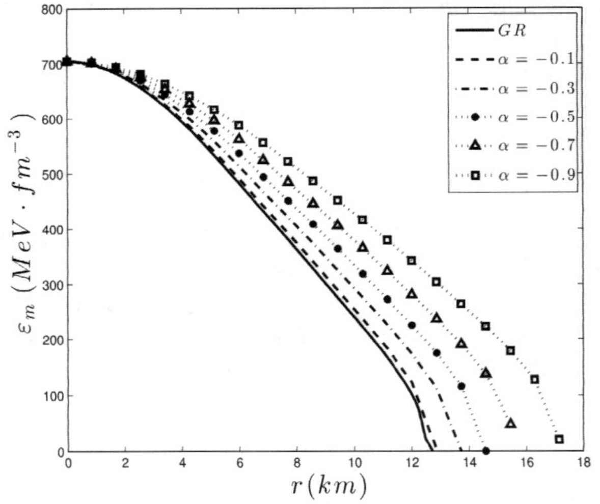

Figure 3. Baryonic energy density profile for different values of the coefficient α. The central Λ-term pressure $p_{\Lambda c}$ has been fixed to $1\varepsilon_0$.

density, depending on the radial distance. One possibility is to model such a dependence but it would be more attractive to derive the dependence of the vacuum fluctuation as a function of the baryonic energy density. This can only be done, up to now, using the semi-classical quantum mechanics, i.e. a fixed metric background. Work in this direction is currently done.

4. Conclusions

In this contribution we applied an extension of the General Relativity, which includes the contribution of dark energy, to neutron stars. At the beginning we shortly resumed the main

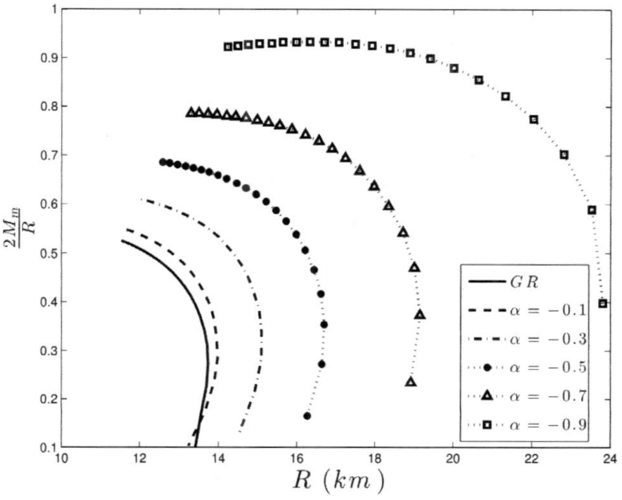

Figure 4. Baryonic compactness vs. total radius for different values of the coefficient α. The central Λ-term pressure $p_{\Lambda c}$ has been fixed to $1\varepsilon_0$.

characteristics of the pc-GR, relevant for the description of pc-neutron stars. Some recent observations of galactic black holes were shown, which indicate a possible inconsistency of GR, namely that the distances deduced from the orbital frequency of QPO and from the measured red-shift are *different*. In contrast, pc-GR can reproduce *the same radial distance*.

The interior region of a neutron star was investigated, including the presence of baryonic and dark energy densities. The equation of state was deduced within the chiral $SU(3)$ model, while the equation of state of the dark energy part is determined through a linear coupling between the dark energy and baryonic density.

Neutron stars with up to six solar masses where obtained, resembling toward larger masses more and more black holes. Larger neutron stars than six solar masses could not be obtained due to the too strong coupling of the energy densities near the surface. Nevertheless, the current study shows that large masses of neutron stars can be obtained and probably no upper limit for stable stars exist.

Details of the pc-GR can be consulted in [2, 3, 9] and the complete investigation on neutron stars will be published elsewhere [12].

5. Acknowledgements
P.O.H. acknowledges financial support from DGAPA (PAPIIT IN103212), CONACyT and the *Frankfurt Institute for Advanced Studies* (FIAS).

References
[1] Dexheimer V and Schramm S 2008 *The Astroph. J.* **3** 943
[2] Hess P O and Greiner W 2009 *Int. J. Mod. Phys.* E **18** 51
[3] Caspar G, Schönenbach T, Hess P and Greiner W 2012 *Int. J. Mod. Phys.* E **20** (S1) 807
[4] Visser M 1996 *Phys. Rev.* D **54** 5116
[5] Barceló C, Liberati S, Sonego S and Visser M 2008 *Phys. Rev.* D **77** 044032
[6] Genzel R, Schödel R, Ott T *et al* 2003 *Nature* **425** 934 (*Preprint* astro-ph/0310821)
[7] Aschenbach B, Grosso N, Porquet D *et al* 2004 *Astronomy and Astrophysics* **417** 71 (*Preprint* astro-ph/0307110)
[8] Iwasawa K, Miniutti G and Fabian A C, 2004 *Month. Notic. of the Roy. Astron. Soc.*, **355**, 1073

Journal of Physics: Conference Series **578** (2015) 012008 doi:10.1088/1742-6596/578/1/012008

[9] Schönenbach T, Caspar G, Hess P O, Boller T, Müller A, Schäfer M and Greiner W 2013 *Month. Notic. of the Roy. Astron. Soc.* **430** 2999
[10] Steiner system XTE J1550-564
[11] Adler R, Bazin M and Schiffer M 1975 *Introduction to General Relativity*, (New York: McGraw Hill)
[12] Rodríguez I, Hess P O, Schramm S and Greiner W, *Neutron Stars within Pseudo-Complex General Relativity* sent for publication.
[13] Ghezzi C R 2011 *Astrophys. Space Sci.* **333** 437
[14] Glendenning N K 2000 *Compact Stars* (Heidelberg: 2nd edition A&A Library, Springer).

Journal of Physics: Conference Series **578** (2015) 012009 doi:10.1088/1742-6596/578/1/012009

A sensitive search for dark energy through chameleon scalar fields using neutron interferometry

W M Snow[1], M Arif[2], B Heacock[3], M Huber[2], K Li[1], D Pushin[4], V Skavysh[1] and A R Young[3]

[1] Indiana University and Center for Exploration of Energy and Matter, Bloomington, Indiana 47408, USA

[2] National Institute of Standards and Technology, Gaithersburg, MD 20899, USA

[3] North Carolina State University, Raleigh, NC 27695, USA

[4] Institute for Quantum Computing and University of Waterloo, Waterloo, ON N2L 3G1, Canada

E-mail: wsnow@indiana.edu

Abstract. The physical origin of the dark energy, which is postulated to cause the accelerated expansion rate of the universe, is one of the major open questions of cosmology. A large subset of theories postulate the existence of a scalar field with a nonlinear coupling to matter chosen so that the effective range and/or strength of the field is greatly suppressed unless the source is placed in vacuum. We describe a measurement using neutron interferometry which can place a stringent upper bound on chameleon fields proposed as a solution to the problem of the origin of dark energy of the universe in the regime with a strongly-nolinear coupling term. In combination with other experiments searching for exotic short-range forces and laser-based measurements, slow neutron experiments are capable of eliminating this and many similar types of scalar-field-based dark energy models by laboratory experiments.

1. Introduction

The observation of an accelerated expansion of the universe was recently recognized with the Nobel Prize in physics [1, 2]. This discovery was a major surprise to physicists and astronomers but has now been confirmed beyond a reasonable doubt [3, 4, 5, 6]. In combination with other cosmological observations it implies that a component of the universe called dark energy constitutes about 70% of the mass-energy of the universe. It is known that a contribution to the vacuum energy density (similar to the cosmological constant that is one of the few allowed terms in classical general relativity) acts like a negative pressure in Einstein's equations and therefore can cause the expansion to accelerate, and the zero-point fluctuations of the quantum fields of the Standard Model provide a natural source for this vacuum energy density. Unfortunately this contribution of the quantum fluctuations of the known Standard Model fields exceeds the observed dark energy density by $50 - 100$ orders of magnitude, thereby constituting the largest finite discrepancy between theory and experiment in the history of physics.

There is no consensus on the nature and physical origin of the dark energy. The number and variety of new theoretical ideas is almost uncountable. At the moment most of the non-theoretical research in this area consists of astronomical observations to more precisely characterize the effects of dark energy on the universe expansion rate and other cosmological and astronomical observables. However an interesting coherent set of theoretical ideas on the origin of dark energy has been proposed which are in principle subject to experimental tests in laboratory measurements. One subset of theories to explain the dark energy postulates the existence of a nonlinear scalar field coupled to matter which propagates over long ranges in vacuum but is highly suppressed in the presence of matter, extending only to distance scales on the order of 100 microns, which is the distance scale associated with the dark energy density.

The first example of this set of ideas [7, 8, 9, 10], known now as the "chameleon mechanism", exploits the $\lambda\phi^4$ term in the usual $L = \frac{1}{2}\partial_\mu\phi\partial^\mu\phi + \frac{m^2}{2}\phi^2 + \lambda\phi^4$ relativistic Lagrangian of a scalar field (such a term must appear in the presence of radiative corrections in quantum field theory even if the original Lagrangian does not include it). The interesting observation made in this model was that, for a broad range of parameters λ, the effective distance that the scalar field propagates is a strong function of the local mass density, and that one can naturally arrange for this distance to be on the order of cosmological scales in the vacuum of outer space relevant for the accelerated expansion observations yet at the same time contract to a very small distance scale in high-density regions (even the mass density of gas at atmospheric pressure is more than sufficient for this suppression). The latter feature allows this chameleon mechanism to escape many precision gravity measurements which have been conducted mainly for other purposes, and therefore this idea along with a range of similar ideas has yet to be ruled out by experiment. The details of these and other similar ideas are under active investigation and have been the subject of a number of recent papers [11, 12, 13, 14, 15, 16, 17, 18, 19, 20, 21] and reviews [22, 23, 24].

2. A Search using Neutron Interferometry

The neutron possesses certain advantages as a system with which to probe for the possible existence of chameleon and related scalar fields in the strongly-coupled chameleon regime. Unlike almost all other experimental probes, the neutron does not possess enough local mass density to suppress the chameleon field but at the same time it can easily penetrate matter densities (typical gas pressures of a few mbar suffice) for which the chameleon field is greatly suppressed. As the chameleon fields must couple directly to mass to be relevant for the observed universe expansion acceleration, other possible probes such as photons possess a more model-dependent coupling to the scalar fields. The chameleon couples to the trace of the energy momentum tensor of matter. In a photon field at the level of the geometrical optics approximation, $T_{\mu\nu} = Ak_\mu k_\nu$ where k_μ is the momentum and A the amplitude, so as long as $k^2 = 0$ as it is for null vectors (photons), we have $T = 0$ and therefore no chameleon effect on photons. This is true in general as long as one does not break conformal invariance and introduce an explicit coupling to F^2 in the Lagrangian, and even in this case the chameleon becomes sensitive to $F^2 = E^2 - B^2$ which is zero for a laser field [25]. So in first approximation chameleons do not see the photon density.

The fact that neutron experiments are a sensitive method to probe chameleon fields was pointed out [27] in the context of measurements of the quantum states of bouncing ultracold neutrons. Recently these measurements have been improved and have set a new experimental limit on chameleon fields [28]. Neutron interferometry measurements can also set interesting limits on chameleon models of dark energy in the "strongly-coupled" limit of the theory. The methodology of the experiments would be very similar to those performed [29] to precisely measure neutron coherent scattering lengths. These measurements are conducted using a gas cell with two chambers, one of which was evacuated and one which contained the gas to be measured. The chameleon field basically fills the volume of the empty cell but in the cell full of gas it would have only extended a distance of order 100 microns outside of the cell walls. Therefore the difference of the phase shifts from these two cells can be analyzed to produce an upper bound on the chameleon field. A calculation [30] showing the limits accessible in neutron interferometry experiments has already appeared.

When combined with other constraints [26] shown in Figure 2 from searches for deviations

Journal of Physics: Conference Series **578** (2015) 012009 doi:10.1088/1742-6596/578/1/012009

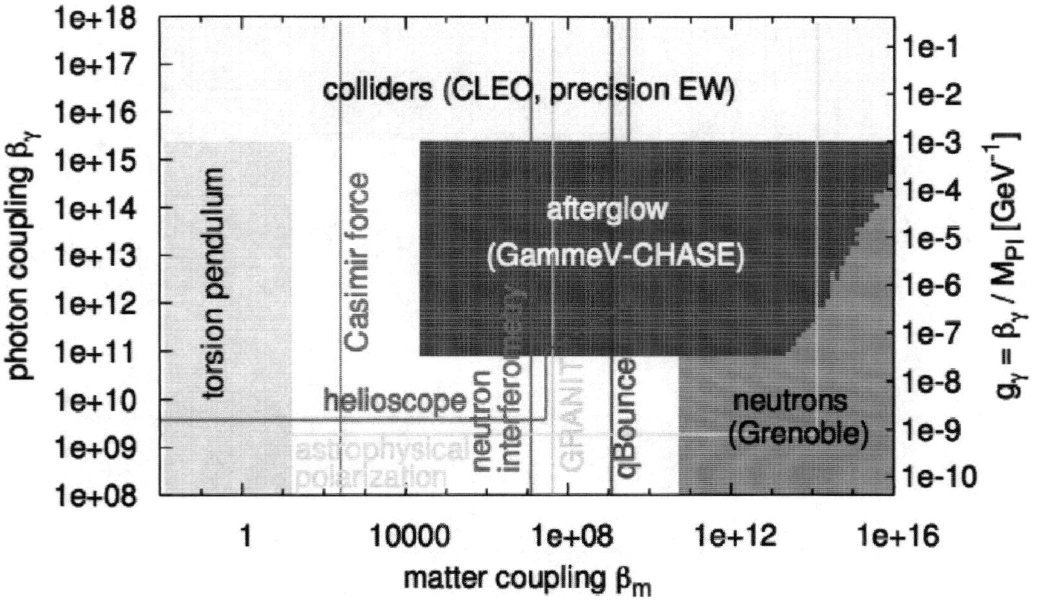

Figure 1. Constraints on the chameleon-matter coupling constant β_m and the chameleon-photon coupling β_γ for a nonlinear effective potential for the chameleon field of the form $V(\phi) = M_\Lambda^4(1 + M_\Lambda/\phi)$ Existing constraints are shown as shaded regions, and forecasts are shown as solid lines. From [26]. The recent results of the qBounce collaboration from [28] now extend slightly beyond the projection shown.

from the gravitational $1/r$ law at short distances at the 100 micron scale (these experiments are sensitive to chameleon fields as they are conducted in high vacuum) and from dedicated experiments involving photons which can couple to the chameleon field indirectly through quantum loop effects, one can foresee the very interesting possibility of experimentally confirming or refuting this model for dark energy. Even if this search finds no evidence for such an interaction it would represent the first nontrivial laboratory test of a plausible explanation for the accelerated expansion of the universe and therefore represent a result of fundamental significance for cosmology. We anticipate that this work might also be used to constrain other dark energy theories based on nonlinear scalar fields. Chameleons are only one example of a larger subclass of scalar field theories which all must possess similar properties to both explain the dark energy and still evade laboratory constraints. With the discovery by the LHC of the first (apparently) fundamental scalar field (the Standard Model Higgs boson) and the continued success of the scalar field mechanism as the explanation for cosmic inflation, past theoretical prejudice against fundamental scalar fields is evaporating with time and perhaps dark energy also really makes use of such a field in nature.

We are preparing a dedicated experiment at the NIST Neutron Interferometer and Optics Facility to search for chameleon fields using an optimized gas cell, a ^{4}He-H2 gas mixture tuned to possess nearly zero strong-interaction-induced phase shift, and a measurement sequence optimized to increase sensitivity to the chameleon field by exploiting its nonlinear contribution to the neutron phase shift as a function of gas pressure. The phase shift expected in the interferometer as a function of gas pressure can be modeled by the equation $\chi = \chi_0 + \chi_{gas} + \chi_{cham}$ where χ_0 is the difference of the phase shifts from the aluminum in the cell walls in the two

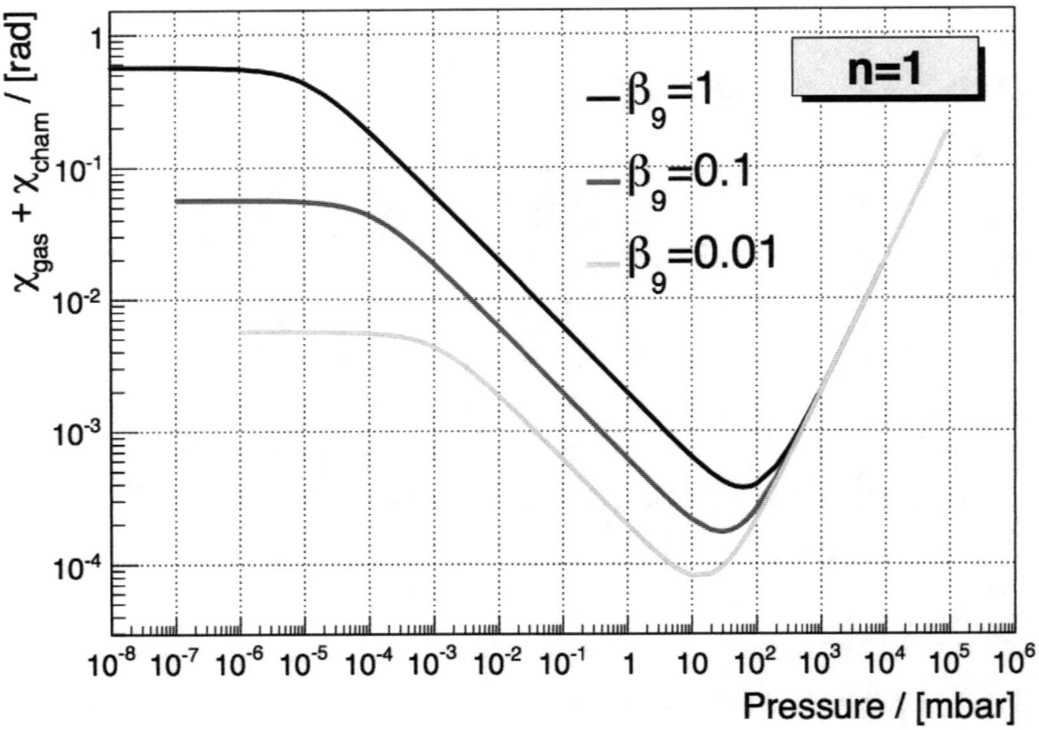

Figure 2. Calculation of the sum of the chameleon and gas mixture phase shifts for the conditions described in the text as a function of the gas pressure. The linear term at high pressure comes from the phase shift due to the neutron optical potential of the gas. The yellow line for β_m of 1×10^7 sets the sensitivity goal for the experiment.

subbeams, $\chi_{gas} = \phi_g P$ is the phase shift from the gas mixture which is linear with pressure, and $\chi_{cham} = \chi_0 f(P)$ is the chameleon phase shift which is a nontrivial function $f(P)$ of pressure which goes to zero at high pressure and approaches unity for pressures around 10^{-4} millibar. Therefore the main idea of the experiment is to conduct a series of measurements of χ as a function of gas pressure and fit to the calculated form of the chameleon field P dependence.

Here we construct a simple model for the experiment to estimate the sensitivity we can achieve. The phase shift from the aluminum cell wall is $\chi_0 = -\lambda N b(D_{II} - D_I)$ where $\lambda = 0.235$ nm is the neutron wavelength, b is the neutron coherent scattering length of aluminum, N is the aluminum density, and D_{II} and D_I are the cell wall thicknesses on the two sides of the interferometer. The world average value for hydrogen is $b_{coh,H2} = -3.741 \pm 0.001$ fm and for ^{4}He is $b_{coh,4He} = +3.26 \pm 0.03$ fm. Assuming we can achieve a gas cell length along the neutron beam of 3 cm with 0.5 cm thick walls which are equal in thickness to the same accuracy achieved in the cell used in the Schoen et al measurement, the phase shift expected from the cell material is about $\chi_0 = 1$ rad and the effect of the gas pressure on the phase shift from the walls of the chamber is negligible. We also expect that we can tune the coherent scattering length of the gas mixture to be zero to an accuracy of 0.006 fm. In this case we expect the cell phase shift from the gas from one chamber of the interferometer with the other side empty to be approximately $\chi_{gas} = 0.002P$ rad with P expressed in units of bar. The neutron phase shift χ_{cham} due to the scalar chameleon field is

$$\chi_{cham} = \int dx \frac{\beta}{M_{PL}} \frac{m^2\phi(x)}{k} \tag{1}$$

where β is the chameleon coupling, M_{PL} is the Planck mass, m is the mass of the neutron, $\phi(x)$ is the chameleon field, and k is the neutron wave vector. The goal of the experiment is to place a limit on β. Recall that the chameleon is a scalar field that is the solution to a nontrivial density-dependent equation of motion engineered to allow the range of the field to constitute the dark energy at the low matter densities relevant for the observations of cosmic acceleration but to "disappear" at normal matter densities beyond the 100 micron dark energy scale. Figure 3 shows the expected total phase shift as a function of gas pressure evaluated for the proposed conditions of the experiment with $\lambda = 0.235$ nm, $L = 3$ cm, $\chi_0 = 1$ rad, and $\chi_{gas} = 0.002P$ rad using our numerical solution of the chameleon equations of motion, which reduce in this case to a nonlinear 1D differential equation. Our calculation is in agreement with that shown by Brax.

A second step that can be taken to further improve the sensitivity of the measurement while still using the existing NIST monolithic interferometer setup would be to exploit the previously-demonstrated possibility of neutron Fabry-Perot cavities and implement one in an arm of the interferometer. An extended series of measurements at the ISIS pulsed neutron facility improved the efficiency of neutron Fabry-Perot cavities made from carefully aligned and separated perfect silicon crystals to the point where neutrons were observed to bounce 2500 times back and forth in the resonator structure [31, 32, 33, 34, 35, 36]. Other experimental measurements of perfect crystal reflectivity with thermal neutrons have observed phenomena which are interpreted as implying that the neutrons made more than 20,000 reflections [37]. Since the chameleon field extends along the entire length of the interferometer arm when the system is in vacuum, the total neutron phase shift from a chameleon field would add coherently as the neutron is reflected back and forth inside the resonator. If one constructs a resonator which operates in both parallel neutron paths of a skew-symmetric interferometer configuration and is able to establish a vacuum in one path and a gas cell in the other path one can increase the sensitivity of the phase shift measurement by an amount as large as the total number of neutron passages through the resonator. However there are major challenges in efficiently implementing this idea on a neutron interferometer. In addition to the obvious issues involved in the stable alignment of separated perfect crystal elements and in loading and unloading the neutrons into the resonator structure, one must also maximize the fraction of neutron phase space accepted by the combination of the neutron interferometer perfect crystal and that accepted in the Bragg geometry of the resonator structure to avoid crippling losses in intensity.

One can imagine yet another step which would further increase the sensitivity of a neutron interferometer-based chameleon search: increase the neutron path length in the interferometer by constructing a separated crystal neutron interferometer. Although such a separated crystal interferometer has not to our knowledge been operated anywhere, it is widely believed by practitioners in the field that it is now technically possible to construct such a device and control it with sufficient precision to maintain a high interferometer contrast [38]. Simulations of the required crystal positioning and operational stability with respect to the Bragg axis show that milliarcsecond precision and mK temperature stability is required for high contrast [39], much more stringent than that needed for a split X-ray interferometer [40, 41]. The milliarcsecond angular stability requirement has already been demonstrated experimentally in the course of measurements conducted at NIST in an attempt to measure the neutron -electron scattering length using neutron interferometry [42]. The crystal positioning can be controlled in principle by registering the appearance of Moire patterns in a position sensitive neutron detector [43]. Were this possible, one could overcome the existing size limitation of monolithic perfect crystal interferometers and achieve a qualitative increase in the neutron path length in the interferometer arms. Development of either effective neutron Fabry-Perot cavities for neutron interferometry or separated perfect crystal interferometers would also be very exciting for many other neutron interferometry studies such as research on decoherence-free subspaces [44]

3. Conclusion

We propose a search for chameleon dark energy fields using the neutron interferometer at NIST. We can perform a sensitive search in the regime of strongly nonlinear coupling and improve the existing upper bound on the coupling parameter β. Further neutron interferometer and gravitational resonance spectroscopy measurements have the capability to either discover chameleons or provide the first laboratory experiment refutation of a plausible dark energy theory. Further improvements in the sensitivity of a neutron interferometer measurement could come in the future in principle from the use of neutron Fabry-Perot cavities [31, 36] or through the implementation of a separated crystal interferometer [38, 42, 43].

4. Acknowledgements

This work was supported by NSF grant PHY-1068712 and by the Indiana University Center for the Exploration of Energy and Matter. M. Snow, K. Li, and V. Skavysh acknowledge the Indiana University Center for Spacetime Symmetries. D. Pushin acknowledges the Canadian Excellence Research, NSERC Discovery and CREATE Programs.

References

[1] Riess A G *et al* 1998 *Astron. J.* **116** 1009
[2] Perlmutter S *et al* 1999 *Astrophys.J.* **517** 565
[3] Komatsu E 2011 *et al Astrophys.J.Suppl.* **192** 18
[4] Larson D *et al* 2011 *Astrophys. J. Suppl.* **192** 16
[5] Suzuki N *et al* 2012 *Astrophys. J.* **746** 85
[6] Sanchez A G *et al* 2012 *Mon. Not. R. Astron. Soc.* **425** 415
[7] Khoury J and Weltman A 2004 *Phys. Rev. Lett.* **93** 171104
[8] Khoury J and Weltman A 2004 *Phys. Rev.* D **69** 044026
[9] Brax P, van de Bruck C, Davis A C, Khoury J, and Weltman A 2004 *Phys. Rev.* D **70** 123518
[10] GubserS S and Khoury J 2004 *Preprint* hep-ph/0405231
[11] Vainshtein A I 1972 *Phys. Lett.* B **39** 393
[12] Dvali G R, Gabadadze G, and Porrati M 2000 *Phys. Lett.* B **485** 208
[13] Deffayet C *et al.* 2002 *Phys.Rev.* D **65** 044026
[14] Olive K A and Pospelov M 2008 *Phys.Rev.* D **77** 043524
[15] Nicolis Rattazzi A R, and Trincherini E 2009 *Phys.Rev.* D **79** 064036
[16] Hinterbichler K and Khoury J 2010 *Phys. Rev. Lett.* **104** 231301
[17] Upadhye A 2012 *Preprint* hep-ph/1209.0211
[18] Upadhye A 2012 *Preprint* hep-ph/1210.7804
[19] Upadhye A, Steven, Chou 2012 *Phys. Rev.* D **86** 035006
[20] Upadhye A, Steven A, Weltman A 2010 *Phys. Rev.* D **81** 015013
[21] Steven A, Upadhye A, Baumbaugh, Chou, Mazur, Tomlin, Weltman, and Wester 2010 *Phys. Rev. Lett.* **105** 261803
[22] Khoury J 2010 *Preprint* astro-ph/1011.5909
[23] Upadhye A 2012 *Preprint* hep-ph/1211.7066
[24] Brax P 2012 *Preprint* hep-ph/1211.5237
[25] P. Brax 2013, private communication
[26] Hansson Adrian P *et al* 2013 *Preprint* hep-ph/1311.0029
[27] Brax P and Pignol G 2011 *Phys. Rev. Lett.* **107** 111301
[28] Jenke T et al 2014 *Phys. Rev. Lett.* **112** 151105
[29] Schoen K, Jacobson D L, Arif M, Huffman P R, Black T C, Snow W M, Lamoreaux S K, Kaiser H, and Werner S A 2003 *Phys. Rev.* C **67** 044005
[30] Brax P, Pignol G, Roulier D 2013 *Phys. Rev.* D 88 083004
[31] Schuster M, Rauch H, Seidl E, and Jericha E 1990 *Phys. Lett.* A **144** 297
[32] Jericha E, Carlile C J, Jaekel M, and Rauch H 1997 *Physica* B **234-236** 1066
[33] Jaekel M R, Carlile C J, Jericha E, Schwab D E, and Rauch H 1999 *Proc. SPIE* **3767** 353

[34] Jericha E, Schwab D E, Jaekel M R, Carlile C J, and Rauch H 2000 *Physica* B **283** 414
[35] Jericha E, Schwab D E, Carlile C J, Jaekel M R, Loidl R, and Rauch H 2000 *Nucl. Instr. Meth.* A **440** 597
[36] Jaekel M R, Jericka E, and Rauch H 2005 *Nucl. Instr. Meth.* A **539** 335
[37] Dombeck T, Ringo R, Koetke D D, Kaiser H, Schoen K, Werner S A, and Dombeck D 2001 *Phys. Rev.* A **64** 053607
[38] Zawisky M, Springer J, Farthofer R, and Kuetgens U 2010 *Nucl. Instr. Meth.* A **612** 338
[39] Springer J, Zawisky M, Farthofer R, Lemmel H, Suda M, and Kuetgens U 2010 *Nucl. Inst. Meth.* A **615** 307
[40] Becker P and Bonse U 1974 *J.Appl.Cryst.* **7**, 593
[41] Liss K D, Hock R, Gomm M, Waibel B, Magerl A, Krisch M, and Tucoulou R 2001 *Proc. SPIE* **4143** 78
[42] Wietfeldt F E, Huber M, Black T C, Kaiser H, Arif M, Jacobson D L, and Werner S A 2005 *Preprint* nucl-ex/0509018
[43] Zawisky M, Springer J, Lemmel H 2011 *Nucl. Inst. Meth.* A. **634** S46
[44] Pushin D A, Huber M G, Arif M, and Cory D G 2011 *Phys. Rev. Lett.* **107** 150401

Journal of Physics: Conference Series **578** (2015) 012010 doi:10.1088/1742-6596/578/1/012010

Dominant Modes in Light Nuclei – *Ab Initio* View of Emergent Symmetries

J P Draayer[1], T Dytrych[1], K D Launey[1], A C Dreyfuss[1], and D Langr[2]

[1]Department of Physics and Astronomy, Louisiana State University,
Baton Rouge, LA 70803, USA
[2] Department of Computer Systems, Czech Technical University in Prague,
Prague, Czech Republic

E-mail: draayer@phys.lsu.edu

Abstract. An innovative symmetry-guided concept is discussed with a focus on emergent symmetry patterns in complex nuclei. In particular, the *ab initio* symmetry-adapted no-core shell model (SA-NCSM), which capitalizes on exact as well as partial symmetries that underpin the structure of nuclei, provides remarkable insight into how simple symmetry patterns emerge in the many-body nuclear dynamics from first principles. This *ab initio* view is complemented by a fully microscopic no-core symplectic shell-model framework (NCSpM), which, in turn, informs key features of the primary physics responsible for the emergent phenomena of large deformation and alpha-cluster substructures in studies of the challenging Hoyle state in Carbon-12 and enhanced collectivity in intermediate-mass nuclei. Furthermore, by recognizing that deformed configurations often dominate the low-energy regime, the SA-NCSM provides a strategy for determining the nature of bound states of nuclei in terms of a relatively small subspace of the symmetry-reorganized complete model space, which opens new domains of nuclei for *ab initio* investigations, namely, the intermediate-mass region, including isotopes of Ne, Mg, and Si.

1. Introduction

The *ab initio* symmetry-adapted no-core shell-model (SA-NCSM), with results that corroborate and are complementary to those enabled within the framework of the no-core shell model (NCSM) [1], and which can be used to facilitate *ab initio* applications to challenging lower *sd*-shell nuclei, reveal that bound states of light nuclei are dominated by high-deformation and low-spin configurations [2]. The applicable symmetries reveal the nature of collectivity in such nuclei and provide a description of bound states in terms of a relatively small fraction of the complete space when the latter is expressed in an $(LS)J$ coupling scheme with the spatial configurations further organized into irreducible representations (irreps) of SU(3). That SU(3) plays a key role tracks with the seminal work of Elliott [3, 4, 5], and is further reinforced by the fact that SU(3) also underpins the microscopic symplectic model [6, 7], which provides a theoretical framework for understanding deformation-dominated collective phenomena in nuclei [7].

While applications to *p*-shell and selected heavier nuclei [2, 9, 10, 11] illustrate the success of the *ab initio* approach, a very simple algebraic interaction, which reduces to the Elliott SU(3) model [3, 4, 5] in the single-shell limit, augmented by the SU(3) symmetry breaking spin-orbit interaction, reproduces characteristic features of the low-lying 0^+ states in ^{12}C as well as ground-

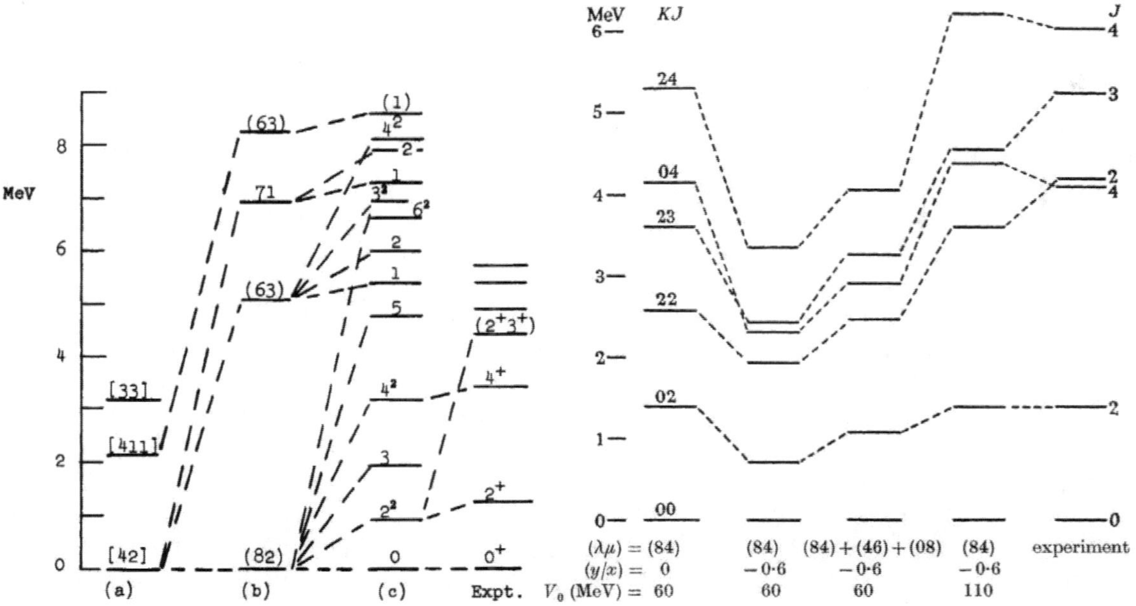

Figure 1. Elliott's SU(3) model applied to *sd*-shell nuclei. Left panel: Spectrum of ^{22}Ne (or ^{22}Mg) (a) with a Majorana potential, (b) with the addition of the second-order SU(3) Casimir invariant, C_2^{su3}, and (c) with the Majorana potential plus an attractive $Q \cdot Q$ interaction [or (b) with the addition of L^2]. Figure taken from [8]. Right panel: Spectrum of ^{24}Mg with a Gaussian central force. Figure taken from [5]. The vertical axis in both figures represents energy in MeV. Note the importance of the most deformed SU(3) configuration (8 2) in ^{22}Ne and (8 4) in ^{24}Mg for reproducing the experimental low-lying states.

state rotational bands in *p* and *sd*-shell nuclei (from Be to Si) [12, 13, 14]. The study of ^{12}C includes the elusive first excited 0_2^+ state, known as the Hoyle state that was predicted based on observed abundances of heavy elements in the universe [15], and which has attracted much recent attention (e.g., see [16, 17]). An implication of the latter is that efforts to reproduce the structure of ^{12}C using a 'bottom up' *ab initio* effective interaction theory may benefit from 'top down' algebraic considerations that serve to expose emergent properties in terms of simple interaction forms that seem to dominate the structure of deformed nuclei, especially the 0^+ states of ^{12}C.

2. Shell models and collectivity-driven models

In the 1950s, two simple models of nuclear structure were advanced that are complementary in nature, namely, the independent-particle model of Mayer and Jensen [18, 19], and the collective model of Bohr and Mottelson [20]. The first of these, which is microscopic in nature, recognizes that nuclei can be described by particles independently moving in a mean field, with the harmonic oscillator (HO) potential being a very good first approximation to the average potential experienced by each nucleon in a nucleus. The second of these, which is collective in nature, recognizes that deformed shapes dominate the dynamics. For example, deformed configurations are found to be important even in a nucleus such as ^{16}O, which is commonly treated as spherical in its ground state, but 20% of the latter is governed by deformed shapes; in addition, the lowest-lying excited 0^+ states in ^{16}O and their rotational bands are dominated by large deformation [21]. Bohr & Mottelson offered a simple but important description of nuclei in terms of the deformation of the nuclear surface and associated vibrations and rotations.

The seminal work of Elliott [3] focused on the key role of SU(3), the exact symmetry of the three-dimensional spherical HO. Within a shell-model framework, Elliott's model utilizes an SU(3)-coupled basis that is related via a unitary transformation to the basis used in the conventional shell model. The new feature here is that SU(3) divides the space into basis states of definite $(\lambda\,\mu)$ quantum numbers of SU(3) linked to the intrinsic quadrupole deformation [22, 23, 24]. E.g., the simplest cases, $(0\,0)$, $(\lambda\,0)$, and $(0\,\mu)$, describe spherical, prolate, and oblate deformation, respectively. For SU(3)-symmetric interactions, the model can be solved analytically. But regardless whether a simple algebraic interaction is used, such as $H = H_{\rm HO} - \frac{\chi}{2} Q \cdot Q$ within a shell (see, e.g., Fig. 1, left), or an SU(3)-symmetry breaking interaction (see, e.g., Fig. 1, right), the results have revealed a striking feature, namely, the dominance of a few most deformed configurations. This has been shown for *sd*-shell nuclei, such as ^{18}Ne, ^{20}Ne, ^{22}Ne, ^{22}Mg, ^{24}Mg, and ^{28}Si, that have been known to possess a clear collective rotational structure in their low-lying states [5, 8]. Furthermore, the pairing interaction, that can be microscopically incorporated into the Elliott model, also breaks the SU(3) symmetry. Nonetheless, it has been shown [25] that this results in mixing of only several different $(\lambda\,\mu)$ configurations and yields outcomes close to experiment and to the energies obtained using full *sd*-shell-model calculations. The SU(3)-symmetry dominance has been also observed in heavier nuclei, where pseudo-spin symmetry and its pseudo-SU(3) complement have been shown to play a similar role in accounting for deformation in the upper *pf* and lower *sdg* shells, and in particular, in strongly deformed nuclei of the rare-earth and actinide regions [26].

Another very significant advance is the microscopic symplectic model [6, 7], developed by Rowe and Rosensteel, which provides a theoretical framework for understanding deformation-dominated collective phenomena in atomic nuclei [7] that involves particle excitations across multiple shells. The significance of the symplectic Sp(3, $\mathbb{R}$) symmetry, the embedding symmetry of SU(3) [Sp(3, $\mathbb{R}$)$\supset$SU(3)], for a microscopic description of a quantum many-body system of interacting particles naturally emerges from the physical relevance of its 21 generators, which are directly related to the particle momentum ($p_{s\alpha}$) and coordinate ($r_{s\alpha}$) operators, with $\alpha = x$, y, and z for the 3 spatial directions and s labeling an individual nucleon, and realize important observables. Namely, Sp(3, $\mathbb{R}$)-preserving operators include: the many-particle kinetic energy $\sum_{s,\alpha} p_{s\alpha}^2/2m$, the HO potential, $\sum_{s,\alpha} m\Omega^2 r_{s\alpha}^2/2$, the mass quadrupole moment $Q_{(2\mathcal{M})} = \sum_s q_{(2\mathcal{M})s} = \sum_s \sqrt{16\pi/5} r_s^2 Y_{(2\mathcal{M})}(\hat{\mathbf{r}}_s)$, and angular momentum L operators, together with multi-shell collective vibrations and vorticity degrees of freedom for a description from irrotational to rigid rotor flows. Indeed, the symplectic Sp(3, $\mathbb{R}$) symmetry underpins the symplectic shell model, which provides a microscopic formulation of the Bohr-Mottelson collective model and is a multiple oscillator shell generalization of the successful Elliott SU(3) model. The symplectic model with Sp(3, $\mathbb{R}$)-preserving interactions[1] has achieved a remarkable reproduction of rotational bands and transition rates without the need for introducing effective charges, while only a single Sp(3, $\mathbb{R}$) configuration is used [7, 28]. A shell-model study in a symplectic basis that allows for mixing of Sp(3, $\mathbb{R}$) configurations due to pairing and non-degenerate single-particle energies above a ^{16}O core [27] has found that using only seven Sp(3, $\mathbb{R}$) configurations is sufficient to achieve a remarkable reproduction of the ^{20}Ne energy spectrum (Fig. 2a) as well as of $E2$ transition rates without effective charges (Fig. 2b).

One of the most successful particle-driven models is the no-core shell model (NCSM), which, in principle, can straightforwardly accommodate any type of inter-nucleon interaction. Specifically, for a general problem, the NCSM adopts the intrinsic non-relativistic nuclear plus Coulomb interaction Hamiltonian defined as follows:

$$H = T_{\rm rel} + V_{NN} + V_{NNN} + \ldots + V_{\rm Coulomb}, \tag{1}$$

[1] An important Sp(3, $\mathbb{R}$)-preserving interaction is $\frac{1}{2}Q \cdot Q = \frac{1}{2}\sum_s q_s \cdot (\sum_t q_t)$, as this realizes the physically relevant interaction of each particle with the total quadrupole moment of the nuclear system.

Journal of Physics: Conference Series **578** (2015) 012010 doi:10.1088/1742-6596/578/1/012010

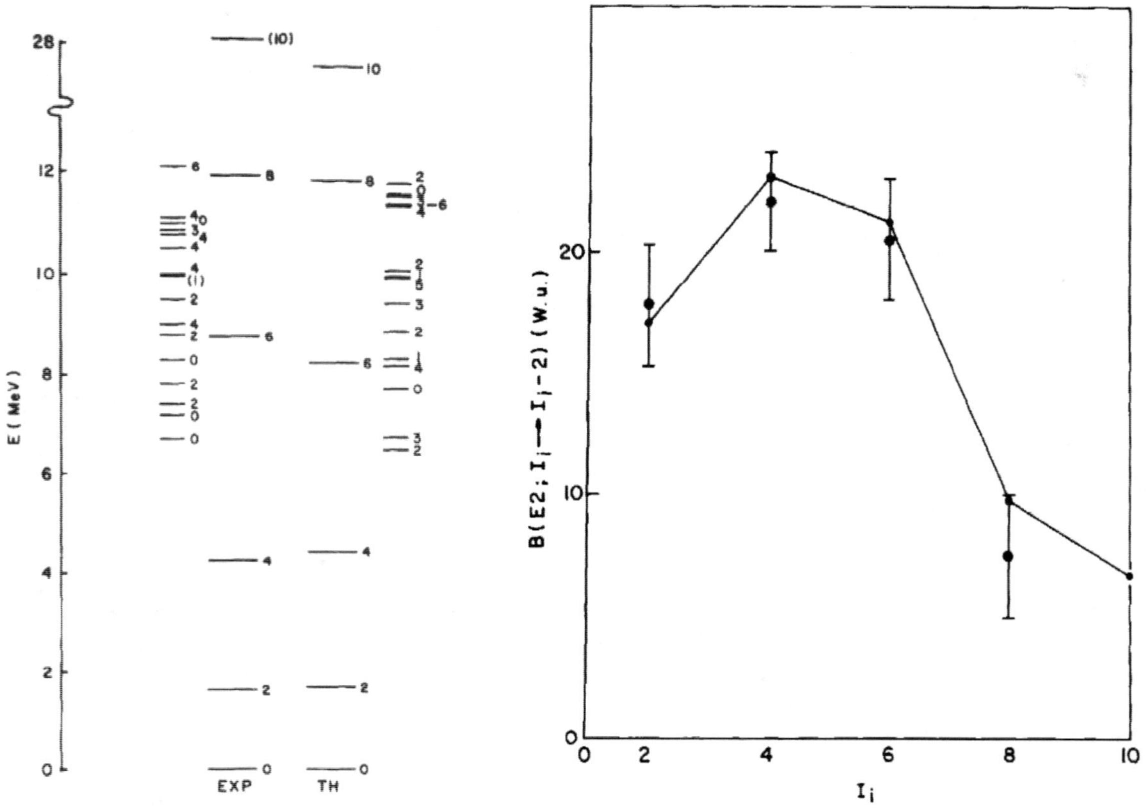

(a) (b)

Figure 2. Microscopic symplectic model with a set of effective single-particle energies, a $Q \cdot Q$-type interaction+pairing for ^{20}Ne. (a) Energy spectrum and (b) $B(E2\downarrow)$ transition strengths for $J^\pi = 2^+, 4^+, 6^+$, and 8^+ (labeled as I_i in the figure), which are calculated *without effective charges* and are found to fall within the uncertainties of the corresponding experimental measurements. Figures taken from [27].

where the V_{NN} nucleon-nucleon and V_{NNN} 3-nucleon interactions are included along with the Coulomb interaction between the protons. The Hamiltonian may include additional terms such as multi-nucleon interactions among more than three nucleons simultaneously and higher-order electromagnetic interactions such as magnetic dipole-dipole terms. It adopts the HO single-particle basis characterized by the $\hbar\Omega$ oscillator strength and retains many-body basis states of a fixed parity, consistent with the Pauli principle, and limited by a many-body basis cutoff $N_{\rm max}$. The $N_{\rm max}$ cutoff is defined as the maximum number of HO quanta allowed in a many-body basis state above the minimum for a given nucleus. It divides the space in "horizontal" HO shells and is dictated by particle-hole excitations (this is complementary to the microscopic symplectic model, which divides the space in vertical slices selected by collectivity-driven rules). It seeks to obtain the lowest few eigenvalues and eigenfunctions of the Hamiltonian (1). The NCSM has achieved remarkable descriptions of low-lying states from the lightest s-shell nuclei up through ^{12}C, ^{16}O, and ^{14}F, and is further augmented by several techniques, such as NCSM/RGM [29], Importance Truncation NCSM [30] and Monte Carlo NCSM [31]. This supports and complements results of other first-principle approaches, such as Green's function Monte Carlo (GFMC) [11], Coupled-cluster (CC) method [32], In-medium SRG [33], and Lattice Effective Field Theory (EFT) [17].

The next-generation *ab initio* symmetry-adapted no-core shell model (SA-NCSM) [2] combines the first-principle concept of the NCSM with symmetry-guided considerations of the collectivity-driven models. For the first time, in the framework of the SA-NCSM, we show the emergence of orderly patterns from first principles [2]. These patterns are linked to the SU(3) and symplectic $Sp(3,\mathbb{R})$ symmetries. This novel feature, in turn, enables the SA-NCSM – by using symmetry-dictated subspaces – to reach new domains of nuclear structure currently inaccessible by *ab initio* calculations, such as isotopes of Ne, Mg, and Si. The model and its recent findings are described in the next section.

Furthermore, also detailed in the next section, a fully microscopic no-core symplectic shell model, NCSpM, is explored to study emergent simple patterns. It utilizes an $Sp(3,\mathbb{R})$-preserving $Q{\cdot}Q$-type interaction plus a symmetry-breaking $l{\cdot}s$ interaction. And while the coupling constants of the two-body interaction we employ can be tied to realistic nucleon-nucleon interactions, we include many-nucleon interactions of a simple form ($e^{-\gamma Q\cdot Q}$) controlled by a parameter γ, the only adjustable parameter in the model. The study has revealed the importance of ultra-large shell-model spaces that are imperative to provide a successful description of the ^{12}C Hoyle state and low-lying states in nuclei from Be to Si [12, 13] (including energy spectra, $E2$ transition strengths, quadrupole moments, and matter rms radii without effective charges).

3. Emergent phenomena

3.1. Ab initio SA-NCSM – mechanism of simple pattern formation from first principles

The *ab initio* symmetry-adapted no-core shell model (SA-NCSM) [2] adopts the first-principle concept and utilizes a many-particle basis that is reduced with respect to the physically relevant $SU(3)\supset SO(3)$ subgroup chain (for a review, see [21]). This allows the full model space to be down-selected to the physically relevant space. The significance of the SU(3) group for a microscopic description of the nuclear collective dynamics can be seen from the fact that it is the symmetry group of the Elliott model [3], and a subgroup of the $Sp(3,\mathbb{R})$ symplectic model [6].

The basis states of the SA-NCSM are based on HO single-particle states and for a given $N_{\max}$, are constructed in the proton-neutron formalism using an efficient construction based on powerful group-theoretical methods. The SA-NCSM basis states are related to the NCSM basis states through a unitary transformation (hence, the SA-NCSM results obtained in a complete $N_{\max}$ space are equivalent to the $N_{\max}$-NCSM results). They are labeled by the $SU(3)\supset SO(3)$ subgroup chain quantum numbers $(\lambda\,\mu)\kappa L$, together with proton, neutron, and total intrinsic spins S_p, S_n, and S. The orbital angular momentum L is coupled with S to the total orbital momentum J and its projection M_J. Each basis state in this scheme is labeled schematically as $|\vec{\gamma}\,(\lambda\,\mu)\kappa L;(S_pS_n)S;JM_J\rangle$. The label κ distinguishes multiple occurrences of the same L value in the parent irrep $(\lambda\,\mu)$, and $\vec{\gamma}$ distinguishes among configurations carrying the same $(\lambda\,\mu)$ and $(S_pS_n)S$ labels.

The *ab initio* SA-NCSM results for p-shell nuclei reveal a dominance of configurations of large deformation in the $0\hbar\Omega$ subspace. For example, the *ab initio* $N_{\max}=8$ SA-NCSM results with the bare N^{3}LO realistic interaction [34] (similarly, for the bare JISP16 realistic interaction [35]) for the 0^+ ground state (*g.st.*) and its rotational band for ^{8}Be reveal the dominance of the $0\hbar\Omega$ component with the foremost contribution coming from the leading $(4\,0)$ $S=0$ irrep (Fig. 3). Furthermore, we find that important SU(3) configurations are then organized into structures with $Sp(3,\mathbb{R})$ symplectic symmetry, that is, the $(4\,0)$ symplectic irrep gives rise to $(6\,0)$, $(4\,1)$, and $(2\,2)$ configurations in the $2\hbar\Omega$ subspace and so on, and those configurations indeed realize the major components of the wavefunction in this subspace. This further confirms the significance of the symplectic symmetry to nuclear dynamics. Similar results are observed for other p-shell nuclei, such as ^{6}Li, ^{6}He, and ^{12}C.

The outcome points to a remarkable feature common to the low-energy structure of nuclei that has heretofore gone unrecognized in other first-principle studies; namely, the emergence, without

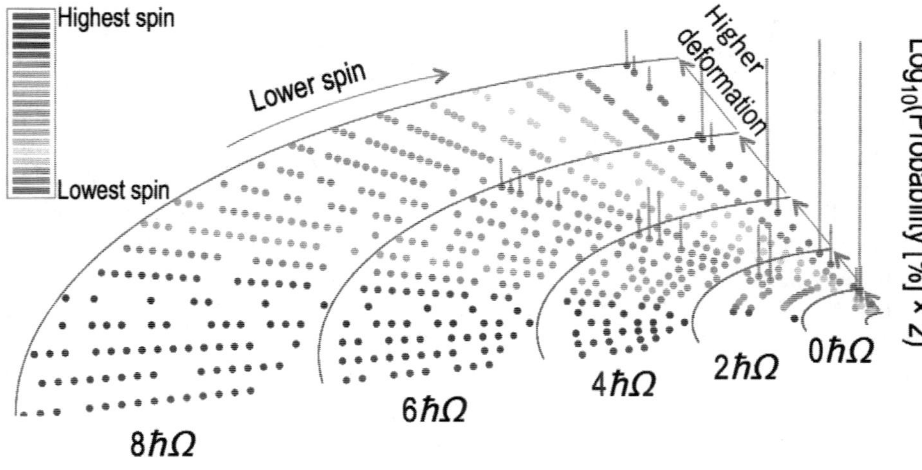

Figure 3. Probability distributions for proton, neutron, and total intrinsic spin components $(S_p S_n S)$ across the Pauli-allowed deformation-related $(\lambda\,\mu)$ values and for $N\hbar\Omega$ subspaces for the calculated 0^+ ground state of ^{8}Be obtained for $N_{\max} = 8$ and $\hbar\Omega = 25$ MeV with the N^{3}LO bare interaction. The concentration of strengths to the far right demonstrates the dominance of collectivity in the calculated eigenstates.

a priori constraints, of simple orderly patterns that favor strongly deformed configurations and low spin values. This feature confirms the dominant role the SU(3) and Sp(3, $\mathbb{R}$) symmetries play in nuclear dynamics and is central to expanding the reach of first-principle studies to heavier nuclei. *Ab initio* investigations of nuclei in the *sd*-shell region are now feasible and the first *ab initio* calculations for ^{24}Si and its mirror nucleus ^{24}Ne (Fig. 3.1) have been achieved in the framework of the SA-NCSM with chiral interactions in an $N_{\max} = 6$ symmetry-winnowed space of 3.5×10^6 dimensionality (compare to the inaccessible space of 8×10^9 dimensionality required for the corresponding conventional NCSM calculations) [36].

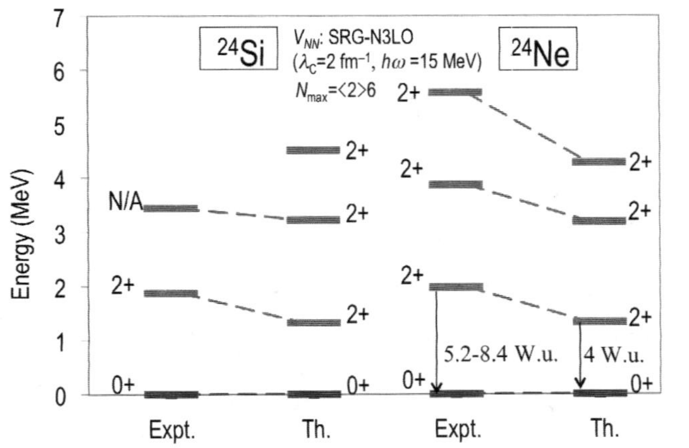

Figure 4. First *ab initio* SA-NCSM calculations for open-shell intermediate-mass nuclei, ^{24}Si and ^{24}Ne, obtained for a selected $N_{\max} = \langle 2 \rangle 6$ model space (that is, the complete space up through $2\hbar\Omega$ and selected large-deformation/low-spin configurations in $4\hbar\Omega$ and $6\hbar\Omega$ subspaces). Calculations are performed for $\hbar\Omega = 15$ MeV and using the SRG-N^{3}LO *NN* interaction for an SRG cutoff $\lambda_c = 2$ fm^{-1}.

In short, the SA-NCSM advances an extensible microscopic framework for studying nuclear structure and reactions that capitalizes on advances being made in *ab initio* methods while exploiting symmetries – exact and partial, known to dominate the dynamics.

3.2. NCSpM model – physics responsible for simple pattern formation

The no-core symplectic shell model (NCSpM) is a fully microscopic no-core shell model that uses a symplectic $Sp(3,\mathbb{R})$ basis and $Sp(3,\mathbb{R})$-preserving interactions. The NCSpM employed within a full model space up through a given $N_{\max}$ coincides with the NCSM for the same $N_{\max}$ cutoff. However, in the case of the NCSpM, the symplectic irreps divide the space into 'vertical slices' that are comprised of basis states of a definite deformation $(\lambda\,\mu)$. Hence, the model space can be reduced to only a few important configurations that are chosen among all possible $Sp(3,\mathbb{R})$ irreps within the $N_{\max}$ model space. The NCSpM, while selecting the most relevant symplectic configurations, is employed to provide shell model calculations beyond current NCSM limits, namely, up through $N_{\max} = 20$ for ^{12}C, the model spaces we found sufficient for the convergence of results [12].

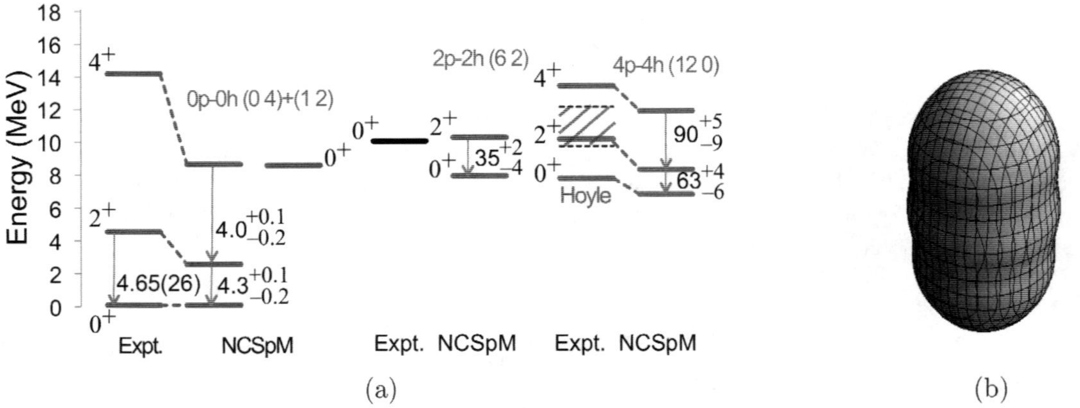

Figure 5. (a) Energy spectra calculated by the NCSpM with $\gamma = -1.71 \times 10^{-4}$ for ^{12}C in an $N_{\max} = 20$ model space and compared to experiment ("Expt."). Figure taken from Ref. [12]. (b) Matter-density profile of the Hoyle state in ^{12}C (with respect to the intrinsic frame) showing the formation of three clusters within a no-core shell-model framework.

We employ a very simple Hamiltonian with an effective interaction derived from the long-range expansion of the two-body central nuclear force together with a spin-orbit term,

$$H_{\text{eff}} = H_0 + \frac{\chi}{2}\frac{\left(e^{-\gamma Q\cdot Q} - 1\right)}{\gamma} - \kappa \sum_{i=1}^{A} l_i \cdot s_i. \tag{2}$$

This includes the spherical HO potential, which together with the kinetic energy yields the HO Hamiltonian, $H_0 = \sum_{i=1}^{A}\left(\frac{\mathbf{p}_i^2}{2m} + \frac{m\Omega^2\mathbf{r}_i^2}{2}\right)$, and the $Q \cdot Q$ quadrupole-quadrupole interaction not restricted to a single shell. For the latter term, the average contribution, $\langle Q\cdot Q\rangle_n$, of $Q \cdot Q$ within a subspace of n HO excitations is removed [37], that is, the trace of $Q \cdot Q$ divided by the space dimension for a fixed n. Hence, the large monopole contribution of the $Q \cdot Q$ interaction is removed, which, in turn, helps eliminate the considerable renormalization of the zero-point energy, while retaining the $Q \cdot Q$-driven behavior of the wavefunctions. This Hamiltonian in its zeroth-order approximation (for parameter $\gamma \to 0$) and for a valence shell goes back to the established Elliott model. We take the coupling constant χ to be proportional to $\hbar\Omega$ and, to leading order, to decrease with the total number of HO excitations, as shown by Rowe [38] based on self-consistent arguments. The success of such an effective nuclear interaction is not unexpected, as the spherical HO potential and the $Q.Q$ interaction directly follow from the second and third term, respectively, in the long-range expansion of any two-body central force, e.g., like the Yukawa radial dependence, $V^{(2)} = \sum_{i<j} V(r_{ij}/a) = \sum_{i<j}(\xi_0 + \xi_2 r_{ij}^2/a^2 + \xi_2 r_{ij}^4/a^4 + \ldots)$ [8], for a range parameter a.

As the interaction and the model space are carefully selected to reflect the most relevant physics, the outcome reveals a quite remarkable agreement with the experiment [12]. The low-lying energy spectrum and eigenstates for ^{12}C were calculated using the NCSpM with H of Eq. (2) for $\hbar\Omega = 18$ MeV given by the empirical estimate $\approx 41/A^{1/3} = 17.9$ MeV and for $\kappa \approx 20/A^{2/3} = 3.8$ MeV (e.g., see [20]). The results are shown for $N_{\max} = 20$, which we found sufficient to yield convergence. This $N_{\max}$ model space is further reduced by selecting the most relevant symplectic irreps, namely, the spin-zero ($S = 0$) $0\hbar\Omega$ 0p-0h (0 4), $2\hbar\Omega$ 2p-2h (6 2), and $4\hbar\Omega$ 4p-4h (12 0) symplectic bandheads together with the $S = 1$ $0\hbar\Omega$ 0p-0h (1 2) and all multiples thereof up through $N_{\max} = 20$ of total dimensionality of 6.6×10^3. In comparison to the experimental energy spectrum (Fig. 5a), the outcome reveals that the lowest 0^+, 2^+, and 4^+ states of the 0p-0h symplectic slices calculated for $\gamma = -1.71 \times 10^{-4}$ closely reproduce the *g.st.* rotational band, while the calculated lowest 0^+ states of the $4\hbar\Omega$ 4p-4h (12 0) and the $2\hbar\Omega$ 2p-2h (6 2) slices are found to lie close to the Hoyle state and the 10-MeV 0^+ resonance (third 0^+ state), respectively. The model successfully reproduces other observables for ^{12}C that are informative of the state structure, such as mass rms radii, electric quadrupole moments and $B(E2)$ transition strengths (Fig. 5a).

A preponderance of the (0 4) $S = 0$ configuration and also (1 2) $S = 1$ configuration is observed for the ground-state rotational band, thereby indicating an oblate shape. The Hoyle-state rotational band includes shapes of even larger deformations but prolate, with the largest contribution of (16 0) (Fig. 5b).

While the model includes an adjustable parameter, γ, this parameter only controls the presence of many-body interactions that cannot be informed by existing realistic NN and NNN interactions. The entire many-body apparatus is fully microscopic and no adjustments are possible. Hence, as γ varies, there is only a small window of possible γ values that, for large enough $N_{\max}$, closely reproduces the relative positions of the three lowest 0^+ states, as well as mass rms radii, electric quadrupole moments and $B(E2)$ transition strengths.

The outcome of the present analysis is not limited to ^{12}C. The model we find is also applicable to the low-lying states of other p-shell nuclei, such as ^{8}Be, as well as sd-shell nuclei without any adjustable parameters [13, 14]. In particular, using the same $\gamma = -1.71 \times 10^{-4}$ as determined for ^{12}C, we describe selected low-lying states in ^{8}Be in an $N_{\max} = 24$ model space with only 3 spin-zero $0\hbar\Omega$ (4 0), $2\hbar\Omega$ (6 0), and $4\hbar\Omega$ (8 0) symplectic irreps. Furthermore, we have successfully applied the NCSpM without any adjustable parameters to the ground-state rotational band of heavier nuclei, such as ^{20}O, ^{20,22,24}Ne, ^{20,22}Mg, and ^{24}Si [14]. This suggests that the fully microscopic NCSpM model has indeed captured an important part of the physics that governs the low-energy nuclear dynamics and informs key features of the interaction and nuclear structure primarily responsible for the formation of such simple patterns.

In short, the NCSpM findings reveal that, to explain emergent simple patterns of deformation and clustering, shell-model spaces well beyond the current limits (up through 22 major HO shells for the Hoyle state) are vital to accommodate particle excitations that appear critical to highly-deformed spatial structures and the convergence of associated observables. The outcome also points to the importance of simple many-body interactions and of the long-range part of the NN interaction, especially through its link to the HO potential and the interaction of individual particles with the total quadruple moment of the nuclear system.

4. Conclusion

Symmetries in atomic nuclei, that have been long recognized, have been recently utilized and further understood in the framework of the *ab initio* symmetry-adapted no-core shell model SA-NCSM as well as of the microscopic no-core symplectic model NCSpM that combine the shell-model and collectivity-driven concepts. The findings pointed to a remarkable new insight, namely, understanding the mechanism on how such simple structures emerge from a fundamental level – the *ab initio* view; but also, understanding the primary physics responsible for the emergent simplicity – the *ab exitu* view.

Furthermore, by utilizing symmetries that are found to underpin nuclear dynamics, the SA-NCSM, while building upon the *ab initio* foundation of the NCSM, has extended its reach to heavier nuclear systems, such as isotopes of Ne, Mg, and Si.

Acknowledgments

We thank the PetaApps Collaboration, in particular, J. P. Vary, P. Maris, U. Catalyurek, E. Saule, and M. Sosonkina, as well as D. Rowe and G. Rosensteel for useful discussions. This work was supported by the U.S. NSF (OCI-0904874), the U.S. DOE (DE-SC0005248), and SURA. ACD acknowledges support by the U.S. NSF (grant 1004822) through the REU Site in Dept. of Physics & Astronomy at LSU. We also acknowledge DOE/NERSC and LSU/LONI for providing HPC resources.

References

[1] Navrátil P, Vary J P and Barrett B R 2000 *Phys. Lett.* **84** 5728
[2] Dytrych T, Launey K D, Draayer J P, Maris P, Vary J P, Saule E, Catalyurek U, Sosonkina M, Langr D and Caprio M A 2013 *Phys. Rev. Lett.* **111** 252501
[3] Elliott J P 1958 *Proc. Roy. Soc. A* **245** 128
[4] Elliott J P 1958 *Proc. Roy. Soc. A* **245** 562
[5] Elliott J P and Harvey M 1962 *Proc. Roy. Soc. A* **272** 557
[6] Rosensteel G and Rowe D J 1977 *Phys. Rev. Lett.* **38** 10
[7] Rowe D J 1985 *Reports on Progr. in Phys.* **48** 1419
[8] Harvey M 1968 *Adv. Nucl. Phys.* **1** 67
[9] Maris P, Shirokov A M and Vary J P 2010 *Phys. Rev. C* **81** 021301(R)
[10] Hagen G, Papenbrock T, Dean D J and Hjorth-Jensen M 2008 *Phys. Rev. Lett.* **101** 092502
[11] Pieper S C, Varga K and Wiringa R B 2002 *Phys. Rev. C* **66** 044310
[12] Dreyfuss A C, Launey K D, Dytrych T, Draayer J P and Bahri C 2013 *Phys. Lett. B* **727** 511
[13] Launey K D, Dytrych T, Draayer J P, Tobin G K, Ferriss M C, Langr D, Dreyfuss A C, Maris P, Vary J P and Bahri C 2013 *Proceedings of the 5th International Conference on Fission and properties of neutron-rich nuclei, ICFN5, November 4 - 10, 2012, Sanibel Island, Florida, edited by J. H. Hamilton and A. V. Ramayya* (World Scientific, Singapore) p 29
[14] Tobin G K, Ferriss M C, Launey K D, Dytrych T, Draayer J P and Bahri C 2014 *Phys. Rev. C* **89** 034312
[15] Hoyle F 1954 *Astrophys. J. Suppl. Ser.* **1** 121
[16] Chernykh M, Feldmeier H, Neff T, von Neumann-Cosel P and Richter A 2007 *Phys. Rev. Lett.* **98** 032501
[17] Epelbaum E, Krebs H, Lee D and Meissner U G 2011 *Phys. Rev. Lett.* **106** 192501
[18] Mayer M G 1949 *Phys. Rev.* **75** 1969
[19] Haxel O, Jensen J H D and Suess H E 1949 *Phys. Rev.* **75** 1766
[20] Bohr A and Mottelson B R 1969 *Benjamin, New York* **1**
[21] Dytrych T, Sviratcheva K D, Draayer J P, Bahri C and Vary J P 2008 *J. Phys. G: Nucl. Part. Phys.* **35** 123101
[22] Rosensteel G and Rowe D J 1977 *Ann. Phys. N.Y.* **104** 134
[23] Leschber Y and Draayer J P 1987 *Phys. Letts. B* **190** 1
[24] Castaños O, Draayer J P and Leschber Y 1988 *Z. Phys. A* **329** 33
[25] Vargas C, Hirsch J and Draayer J 2001 *Nucl. Phys. A* **690** 409
[26] Bahri C, Draayer J and Moszkowski S 1992 *Phys. Rev. Lett.* **68** 2133
[27] Draayer J, Weeks K and Rosensteel G 1984 *Nucl. Phys.* **A419** 1
[28] Bahri C and Rowe D J 2000 *Nucl. Phys. A* **662** 125
[29] Navrátil P, Quaglioni S, Stetcu I and Barrett B R 2009 *J. Phys. G: Nucl. Part.* **36** 083101
[30] Roth R and Navrátil P 2007 *Phys. Rev. Lett.* **99** 092501
[31] Abe T, Maris P, Otsuka T, Shimizu N, Utsuno Y and Vary J 2012 *Phys. Rev. C* **86** 054301
[32] Wloch M, Dean D J, Gour J R, Hjorth-Jensen M, Kowalski K, Papenbrock T and Piecuch P 2005 *Phys. Rev. Lett.* **94** 212501
[33] Tsukiyama K, Bogner S K and Schwenk A 2011 *Phys. Rev. Lett.* **106** 222502

[34] Entem D R and Machleidt R 2003 *Phys. Rev. C* **68** 041001
[35] Shirokov A, Vary J, Mazur A and Weber T 2007 *Phys. Lett. B* **644** 33
[36] Dytrych T, Launey K D and Draayer J P 2014 *McGraw-Hill Yearbook of Science and Technology* YB140314
[37] Rosensteel G and Draayer J P 1985 *Nucl. Phys. A* **436** 445
[38] Rowe D J 1967 *Phys. Rev.* **162** 866

Recent advances in exact solutions of pairing models

S Pittel

Bartol Research Institute and Department of Physics and Astronomy, University of Delaware, Newark, Delaware 19716, USA
Instituto de Ciencias Nucleares, Universidad Nacional Autónoma de México, Aptdo. Postal 70-543, 04510 México, D.F., México

E-mail: pittel@bartol.udel.edu

Abstract. Following a brief overview on the subject of Exactly Solvable Pairing Models, we describe two recent developments in the field that could have a future impact in nuclear structure theory. One concerns a recent extension to include the continuum and the second concerns development of the hyberbolic pairing model as an exactly solvable approximation to Gogny pairing.

1. Introduction

In recent years, there has been great interest in the exact solution of the BCS pairing hamiltonian. Originally discovered by Richardson in the early 60s, this subject was rediscovered some 40 years later in the context of ultrasmall superconducting grains [1], to appropriately describe the crossover from superconductivity to a normal metal as a function of the grain size. Since then, there has been a flurry of activity in the field, from the extension of the Richardson exact solution to several families of exactly-solvable models, now called the Richardson-Gaudin (RG) models [2], to the application of these models in many different areas of quantum many-body physics. The current status of the subject, with particular emphasis on its role in nuclear structure physics, was recently reviewed [3] as a chapter of the book "Fifty Years of Nuclear BCS". Here we focus on two recent developments in the field, developments that could perhaps prove useful in the ongoing program to develop a unified microscopic theory of finite nuclei.

The outline of the presentation is as follows. In Section 2, we briefly review Richardson's solution of the BCS pairing hamiltonian and then in Section 3 discuss its extension by Id Betan [4, 5] to properly include the continuum. In Section 4, we briefly describe the two families of Richardson-Gaudin models associated with the SU(2) algebra and then in Section 5 focus on a recent application [6] of the hyperbolic model as an exactly solvable approximation to Gogny pairing. Finally, Section 6 provides a brief summary of the key points of the presentation.

2. Richardson's solution of the BCS pairing hamiltonian

We focus on a pairing hamiltonian with constant strength G acting in a space of doubly-degenerate time-reversed states $(k, \bar{k})$,

$$H_P = \sum_k \varepsilon_k c_k^\dagger c_k - G \sum_{k,k'} c_k^\dagger c_{\bar{k}}^\dagger c_{\bar{k}'} c_{k'} \ , \qquad (1)$$

where ε_k are the single-particle energies for the doubly-degenerate orbits $k, \bar{k}$.

Cooper [7] considered the addition of a pair of fermions with an attractive pairing interaction on top of an inert Fermi sea (FS) under the influence of this hamiltonian, showing that the eigenstate is

$$|\Psi_{Cooper}\rangle = \sum_{k>k_F} \frac{1}{2\varepsilon_k - E} c_k^\dagger c_{\bar{k}}^\dagger |FS\rangle \ , \tag{2}$$

with E the energy eigenvalue.

Richardson [8, 9] proposed an ansatz for the exact solution of the hamiltonian (1) based closely on Cooper's original idea. For a system with $2M + \nu$ particles, with ν of them unpaired, his ansatz involved the state

$$|\Psi\rangle = B_1^\dagger B_2^\dagger \cdots B_M^\dagger |\nu\rangle \ , \tag{3}$$

with the collective pair operators $B_\alpha^\dagger$ that build this state having the form found by Cooper for the one-pair problem,

$$B_\alpha^\dagger = \sum_{l=1}^{L} \frac{1}{2\varepsilon_k - E_\alpha} c_k^\dagger c_{\bar{k}}^\dagger \ . \tag{4}$$

Here L is the number of doubly-degenerate single-particle levels and

$$|\nu\rangle \equiv |\nu_1, \nu_2 \cdots, \nu_L\rangle \tag{5}$$

is a state of ν unpaired fermions ($\nu = \sum_k \nu_k$, with $\nu_k = 1 \ or \ 0$), defined by $c_k c_{\bar{k}} |\nu\rangle = 0$, and $n_k |\nu\rangle = \nu_k |\nu\rangle$.

In the one-pair problem, the quantity E_1 entering (4) is the eigenvalue of the pairing hamiltonian, as shown by Cooper. In the M-pair problem, Richardson proposed to use the M quantities E_α (called the pair energies) as parameters chosen to fulfill (if possible) the eigenvalue equation $H_P |\Psi\rangle = E |\Psi\rangle$. He showed that it is indeed an exact eigenstate of the pairing hamiltonian if these pair energies satisfy a set of M non-linear coupled equations

$$1 - \frac{G}{2} \sum_{k=1}^{L} \frac{2 - 2\nu_k}{2\varepsilon_k - E_\alpha} - 2G \sum_{\beta(\neq\alpha)=1}^{M} \frac{1}{E_\beta - E_\alpha} = 0 \ , \tag{6}$$

which are now called the Richardson equations. The second term represents the interaction between particles in a given pair, whereas the third term represents the interaction between pairs. The eigenvalues of H associated with a given set of pair energies emerging from (6) are given by

$$E = \sum_{k=1}^{L} \varepsilon_k \nu_k + \sum_{\alpha=1}^{M} E_\alpha \ , \tag{7}$$

namely as a sum of the pair energies plus a contribution from the unpaired particles.

Each independent solution of the set of Richardson equations defines a set of M pair energies that completely defines a particular eigenstate (3, 4). All eigenstates of the pairing hamiltonian can be obtained in this way, both for systems with an odd or an even number of particles. The ground eigenstate is the energetically lowest solution in the $\nu = 0$ or $\nu = 1$ sector, depending on whether the system has an even or an odd number of particles, respectively.

There are few points worth noting here. First, if one of the pair energies E_α is complex, then its complex-conjugate E_α^* is also a solution, as required for $|\Psi\rangle$ to preserve the time-reversal invariance of the hamiltonian. Second, from the structure of the Richardson pair (4), we see that a pair energy that is close to the energy of an particular unperturbed pair $2\varepsilon_k$ is dominated

Journal of Physics: Conference Series **578** (2015) 012011 doi:10.1088/1742-6596/578/1/012011

by this configuration, so that the associated pair is uncorrelated. In contrast, if a pair energy lies sufficiently far from any $2\varepsilon_k$ in the complex plane, the resulting pair will be highly correlated. Lastly, the set of coupled non-linear Richardson equations can be solved numerically, and efficient algorithms for doing so have been developed (see *e.g.* [10]).

3. Extension to the continuum

Richardson's solution focused on pairing over a set of discrete states. In nuclear physics, a realistic single-particle spectrum includes both discrete states and continuum states. The latter are especially important when they are sufficiently close to the Fermi energy such that pair scattering into them can be important or when the Fermi level is itself in the continuum. This is the case both for weakly-bound systems and for unbound systems. A description of such systems requires explicit treatment of pair scattering into the continuum.

The first effort to include pair scattering to the continuum within a Richardson approach was reported by Hasegawa and Kaneko [11]. In that work, however, only the effect of resonances was considered. As a consequence, their calculations produced complex energies even for bound states of the system.

As is well known, a proper treatment of the continuum should treat not only resonances but also the background states obtained from contours in the complex plane that enclose all the resonances included. The first work that treated the full continuum was reported recently by Id Betan in two papers [4, 5]. Here we briefly review the key points of this work and summarize the key results.

In the presence of the continuum, Richardson's equation (6) becomes

$$1-\frac{G}{2}\sum_{k=1}^{L}\frac{d_k}{2\varepsilon_k-E_\alpha}-\frac{G}{2}\int_0^\infty\frac{g(\varepsilon)}{\varepsilon-E_\alpha}\,d(\epsilon)-2G\sum_{\beta(\neq\alpha)=1}^{M}\frac{1}{E_\beta-E_\alpha}=0\ ,t \tag{8}$$

where d_k is the occupation of level k, including blocking effects, and $g(\epsilon)$ is the Continuum Single Particle Level Density (CSPLD).

In a proper treatment of the continuum, the CSPLD should include contributions both from resonances (res) and from background (bkgrd) states. In contrast to the earlier work of ref. [11], Id Betan included both contributions, *viz.*

$$g(\epsilon)=g_{res}(\epsilon)+g_{bkgrd}(\epsilon) \tag{9}$$

The resonant contributions can be modeled in the usual Breit-Wigner form. The non-resonant (background) can be treated by rotating the integration contour of the resonant part to the imaginary axis and then using the Cauchy Theorem. More details can be found in ref. ([5].

The most recent paper [5] used this formalism to treat a nuclear chain that includes both bound and unbound systems, namely the even-A Carbon isotopes up to ^{28}C. When the system is bound, the pair energies that contribute to the ground state emerged in complex conjugate pairs, thus preserving the real nature of the ground state energy. This can be seen clearly from Figure 3 of ref. [5], which gives the pair energies for ^{22}C, the last of the bound even-C isotopes.

Once the system becomes unbound, however, this ceases to be the case. Now the pair energies that contribute to the ground state do not occur in complex conjugate pairs. More specifically, as can be seen from figure 11 of Ref. [5], where the pair energies for ^{28}C are displayed, there are a series of pair energies for which the real part is negative and these occur in nearly complex conjugate pairs. These are indeed the pair energies that are analogous to those for bound ^{22}C. The remaining pair energies all have positive real parts and are typically far from being in complex conjugate pairs. From these results, we see how complex energies and thus widths arise for the states of an unbound system within a Richardson approach that includes the true continuum.

Journal of Physics: Conference Series **578** (2015) 012011 doi:10.1088/1742-6596/578/1/012011

4. Generalization to the Richardson-Gaudin class of integrable models

Here we discuss how to generalize the exactly-solvable BCS pairing model to a wider variety of exactly-solvable models, the so-called Richardson-Gaudin models [12], all of which are based on the $SU(2)$ algebra.

We begin by introducing the generators of $SU(2)$,

$$K_j^0 = \frac{1}{2}\left(\sum_m a_{jm}^\dagger a_{jm} - \Omega_j\right), \quad K_j^+ = \sum_m a_{jm}^\dagger a_{j\bar{m}}^\dagger, \; K_j^- = (K_j^+)^\dagger. \tag{10}$$

Here $a_{jm}^\dagger$ creates a fermion in single-particle state jm, $j\bar{m}$ is the time reverse of jm, and $\Omega_j = j + \frac{1}{2}$ is the pair degeneracy of orbit j. These operators fulfill the usual $SU(2)$ commutation algebra $[K_j^+, K_{j'}^-] = 2\delta_{jj'}K_j^0$, $[K_j^0, K_{j'}^\pm] = \pm\delta_{jj'}K_j^\pm$.

We next consider the most general set of L Hermitian and number-conserving operators that can be built up from the generators of $SU(2)$ for each level j with linear and quadratic terms only,

$$R_i = K_i^0 + 2g\sum_{j(\neq i)}\left[\frac{X_{ij}}{2}\left(K_i^+K_j^- + K_i^-K_j^+\right) + Y_{ij}K_i^0K_j^0\right] . \tag{11}$$

It turns out that there are essentially two sets of conditions on the matrices X and Y under which the set of R operators commute and are complete. The two families of solutions associated with these conditions are referred to as the rational and hyperbolic families, respectively.

i. The rational family

$$X_{ij} = Y_{ij} = \frac{1}{\eta_i - \eta_j} \tag{12}$$

ii. The hyberbolic family

$$X_{ij} = 2\frac{\sqrt{\eta_i\eta_j}}{\eta_i - \eta_j} \; , \; Y_{ij} = \frac{\eta_i + \eta_j}{\eta_i - \eta_j} \tag{13}$$

In both families, the parameters η_i are a set of L free real numbers.

A third family, called the trigonometric, was shown to be equivalent to the hyperbolic family in [12], as you can pass from the hyperbolic to the trigonometric family by simply replacing real etas by imaginary etas.

The traditional pairing model is an example of the rational family, as it can be obtained as a linear combination of the integrals of motion, $H_P = \sum_j \varepsilon_j R_j(\varepsilon_j)$, with $\eta_j = \varepsilon_j$.

But it is not the only one.

Any hamiltonian that can be expressed as a linear combination of the R_i operators, whether for the rational family or for the hyperbolic family, can be solved exactly using the same methods as were used by Richardson for the pure pairing model. In all cases, one is led to a system of coupled non-linear equations, whose solutions can be used to generate all eigensolutions of that integrable hamiltonian.

5. Application of the hyperbolic model in nuclear physics

We now discuss the one application in nuclear physics reported for the hyperbolic class of SU(2) RG models [6]. What was shown in this work is that hyperbolic model gives rise to a separable pairing hamiltonian with two free parameters, which after appropriate choice gives a good approximation to Gogny pairing, *but which can be can be solved exactly.*

The hamiltonian for the hyperbolic model we will discuss can be obtained as a simple linear combination of the hyperbolic integrals of motion, $H = \lambda \sum_i \eta_i R_i$. By defining $\lambda =$

$[1 + 2\gamma(1 - M) + \gamma L_C]^{-1}$ and carrying out some algebraic manipulations, we end up with a hamiltonian

$$H = \sum_i \eta_i K_i^0 \ - \ G \sum_{i,j} \sqrt{\eta_i \ \eta_j} \ K_i^+ K_j^- \ , \tag{14}$$

where $G = 2\lambda\gamma$ is a free parameter.

If we define $\eta_i = 2(\varepsilon_i - \alpha)$, where ε_i is the single-particle energy of level i and α plays the role of an interaction cutoff and if we make use of the pair representation of the $SU(2)$ generators, we are led to a hamiltonian of the form

$$H = \sum_i \varepsilon_i \left(c_i^\dagger c_i + c_{\bar{i}}^\dagger c_{\bar{i}} \right) - 2G \sum_{i,j} \sqrt{(\alpha - \epsilon_i)} \sqrt{(\alpha - \epsilon_j)} \ c_i^\dagger c_{\bar{i}}^\dagger c_{\bar{j}} c_j \ . \tag{15}$$

By introducing the interaction cutoff α, we are led to a separable and exactly-solvable pairing hamiltonian with state-dependent pairing strengths in which the strengths do not grow unphysically as we increase the energy.

This separable pairing hamiltonian can be solved exactly using the RG approach. The resulting energies are given by

$$E = 2\alpha M + \sum_i \varepsilon_i \nu_i + \sum_\beta E_\beta \tag{16}$$

where the pair energies E_β are solutions of the set of non-linear coupled Richardson equations

$$\frac{1}{2} \sum_i \frac{1}{\eta_i - E_\beta} - \sum_{\beta'(\neq\beta)} \frac{1}{E_{\beta'} - E_\beta} = \frac{Q}{E_\beta} \ , \tag{17}$$

where $Q = \frac{1}{2G} - \frac{L}{2} + M - 1$. Each solution of Eq. (17) defines a unique eigenstate of the hyperbolic hamiltonian (15).

To see whether this model hamiltonian is useful, the authors of Ref. [6] applied it to two well-deformed nuclei, ^{238}U and ^{154}Sm. Here we discuss their results and conclusions for ^{238}U.

To carry out analysis, they first took the single-particle energies ϵ_i from a Gogny HFB calculation for this nucleus. and then, using a BCS treatment of the hyperbolic hamiltonian (straightforward because of the separable character of the interaction), fitted the two parameters α and G to the gaps and pairing tensors of the Gogny HFB calculation. An optimal fit was obtained with the choice of parameters $\alpha = 25.25 \ MeV$ and $G = 2 \times 10^{-3} \ MeV$. Comparison of the results of the Gogny HFB calculation and the BCS results for the hyperbolic hamiltonian with this choice of parameters is shown in Fig. 1. Note that the hyperbolic model reproduces Gogny's fall off in Δ_i with increasing energy, in contrast to constant-G pairing for which Δ_i would remain constant. Overall, the hyperbolic model when treated in mean field gives a very good description of the corresponding mean-field physics of Gogny.

What is perhaps more important, however, is that this hamiltonian is exactly solvable, so that it can model in an approximate way the physics of Gogny beyond mean-field. This too was studied in Ref. [6]. The calculations were carried out in an energy window that extended 30 MeV above and below the Fermi surface. For such a window of energies, the full space involves 148 single-particle levels for the 46 active proton pairs, for which the exact shell-model space contains 4.8×10^{38} states. Diagonalization in such a space is clearly prohibitive. In contrast, the exact solution of the hyperbolic hamiltonian requires the quite tractable solution of 46 coupled non-linear equations. When these calculations were carried out, it was found that the exact solution of the hyperbolic hamiltonian gives roughly 2 MeV more correlation energy than the BCS solution, suggesting the importance of going beyond mean field. As noted earlier, calculations were also carried out for ^{154}Sm and can in principle be carried out systematically to treat pairing in heavy nuclei beyond mean field.

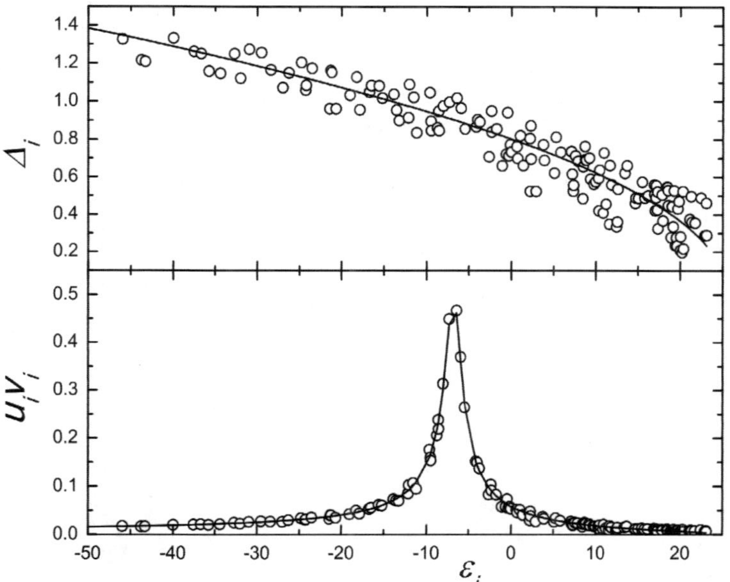

Figure 1. Gaps Δ_i and pairing tensor $u_i v_i$ for protons in ^{238}U. Open circles are Gogny HFB results in MeV. Solid lines are BCS results of the hyperbolic hamiltonian in MeV.

6. Summary and concluding remarks

Following a brief review of Richardson's exact solution of the pairing model and its generalization to a wider class of exactly-solvable RG models, we discussed two recent advances that have the potential of impacting current efforts to develop a unified microscopic theory of atomic nuclei. One was the recent extension by Id Betan of Richardson's solution to the continuum, where we showed that when the continuum is treated appropriately the proper behavior is achieved on transitioning from bound to unbound systems. This could prove useful in the description of very weakly bound and unbound nuclear systems. The other was a first application of the hyperbolic model to nuclear systems, where it was shown that with appropriate choice of the parameters of the model it provides a very good and exactly-solvable approximation to Gogny pairing. This could prove useful in systematic HF + exact pairing calculations of heavy nuclei.

Acknowledgments

The work reported here is based in part on a review article written by the author with Jorge Dukelsky, whose contributions are greatly appreciated and gratefully acknowledged. The work was supported in part by the National Science Foundation under grant # PHY-0854873, CONACYT México and la Academia Mexicana de Ciencias.

References

[1] Sierra G, Dukelsky J, Dussel G G, von Delft J and Braun F 2000 *Phys. Rev.* B **61** 11890
[2] Dukelsky J, Pittel S, and Sierra G 2004 *Rev. Mod. Phys.* **76** 643
[3] Dukelsky J and Pittel S 2013 *Fifty Years of Nuclear BCS* eds R Broglia and V Zelevinsky (Singapore: World Scientific Publishing Company) p 200
[4] Id Betan R 2012 *Nucl. Phys.* A **879** 14
[5] Id Betan R 2012 *Phys. Rev.* C **85** 064309

[6] Dukelsky J, Lerma H S, Robledo L M, Rodriguez-Guzman R and Rombouts S M A 2011 *Phys. Rev.* C **84** 061301
[7] Cooper L N 1956 *Phys. Rev.* **104** 1189
[8] Richardson R W 1963 *Phys. Lett.* **3** 277
[9] Richardson R W 1964 *Nucl. Phys.* **52** 221
[10] Dussel G G, Pittel S, Dukelsky J, Sarriguren P 2007 *Phys. Rev.* C **76** 011302
[11] Hasegawa M and Kaneko K 2003 Effects of resonant single-particle states on pairing correlations *Phys. Rev.* C **67** 024304
[12] Dukelsky J, Esebbag C and Schuck P 2001 Class of Exactly Solvable Pairing Models *Phys. Rev. Lett.* **87** 066403

Journal of Physics: Conference Series **578** (2015) 012012 doi:10.1088/1742-6596/578/1/012012

Inter-band B(E2) transitions strengths in $^{160-170}$Dy nuclei

Carlos E Vargas[1], **Víctor Velázquez**[2] **and Sergio Lerma**[1]

[1] Facultad de Física e Inteligencia Artificial, Universidad Veracruzana. Lomas del Estadio S/N; Zona UV, Xalapa, Ver., 91000, México

[2] Facultad de Ciencias, Universidad Nacional Autónoma de México, Apartado Postal 70-542, 04510 México D.F., México

E-mail: `cavargas@uv.mx`

Abstract. The rare earth region of the nuclear landscape is characterized by a large collectivity observed. The microscopic studies are difficult to perform in the region due to the enormous size of the valence spaces. The use of symmetries based models avoids that problem, because the symmetry allows to choose the most relevant degrees of freedom for the system under consideration. We present theoretical results for electromagnetic properties in $^{160-168}$Dy isotopes employing the pseudo-SU(3) model. In particular, we study the B(E2) inter-band transition strengths between the ground state, γ and β-bands. The model succesfully describes in a systematic way rotational features in these nuclei and allows to extrapolate toward the midshell nucleus ^{170}Dy.

1. Introduction

The rare-earth region of the nuclear landscape provides an excelent opportunity to study unique aspects of nuclear structure around midshell region as new modes of excitation and collectivity, the role of single-particle levels [1] and even possible changes in the shell structure. Nevertheless, despite the recent progress in experimental techniques, the rare-earth neutron-rich nuclei remains a region of the nuclear chart which has been seldom studied. For example, in heavy-ion induced multinucleon transfer reaction, the combination of stable projectiles and stable targets has unfavorable kinematic matching conditions leading to heavy neutron-rich nuclei, resulting in small cross sections that limits its use for populating neutron-rich nuclei [2].

Many low-lying collective nuclear properties are expected to appear in this region, because there nuclei have large numbers of valence protons and neutrons. The neutron-rich nuclei Er, Yb and Hf exhibit a minimum in the 2^+ energies and a maximum of deformation at N=104 [3]. Assuming standard spherical shell gaps, ^{170}Dy has the larger number of valence particles of any nucleus lighter than the doubly magic lead, and hence it is expected to be the most collective nucleus in the region [4, 5, 6, 7]. Experimental works in the lighter dysprosium isotopes have shown that there is an enhancement of the collectivity in ^{164}Dy, a relative flatness of the moment of inertia in ^{166}Dy, and ^{168}Dy shows again an apparent enhancement of deformation [8]. Nevertheless, to date the energies of the levels in ^{170}Dy have not been measured, and hence its degree of collectivity is still unknown. In addition, the known saturation of B(E2) in the region [5, 6, 7] and the decrease in moment of inertia between N=98-100 in the gadolinium isotopes [9] open the following question of the collectivity in the region: is ^{170}Dy indeed the most collective of the region or is a little of the collectivity lost when we reach N=104?.

There are very different channels for producing the dysprosium neutron-rich nuclei. Between these, C. Wu and coworkers have used quasielastic two-neutron transfer reaction to study the neutron-rich ^{166}Dy [2], two-proton pick-up reaction has allowed to populate ^{168}Dy [8, 10], and projectile fragmentation of ^{208}Pb was used [11, 12] to synthetize the midshell ^{170}Dy.

Regarding the use of theoretical schemes, the large valence spaces associated with heavy-nuclei have implied a slow progress, limiting the number of models capable of dealing with ^{170}Dy. A shell model description of heavy nuclei requires assumptions that include a systematic and proper truncation of the Hilbert space. The symmetry based SU(3) shell model [13, 14] has been successfully applied in light nuclei, where a harmonic oscillator mean-field and a residual quadrupole-quadrupole interaction can be used to describe dominant features of the nuclear spectra. However, the strong spin-orbit interaction renders the SU(3) model useless in heavier nuclei, while at the same time pseudo-spin emerges as a good symmetry [15, 16]. This symmetry has its origin in the relativistic mean field for heavy-nuclei [17, 18]. The consistency of this symmetry has determined the success of the pseudo-SU(3) model [19].

Four decades have passed since the pseudo-SU(3) model was first introduced. During that time period, and especially over the last few years, the pseudo-SU(3) model has been used to describe a wide variety of very different nuclear phenomena. Initially it was considered the pseudo-spin as a dynamical symmetry [20]. With the development of a computer code it has been possible to calculate reduced matrix elements of physical operators (i.e. the breaking-symmetry Nilsson single-particle energies and pairing correlations) between different SU(3) irreps [21]. This enabling technology allows, among others, fully microscopic studies of energy levels belonging to normal parity bands, BE2 intra and interband transitions and the analysis of the interplay between orbital and spin components of the scissors mode [22, 23, 24, 25] in both, even-even and A-odd nuclei. The high density of 0^+ states in ^{158}Gd has also been described with this model [26]. Despite this long history, however, the model has been rarely used to predict the phenomenology in nuclei far from the stability region. Nevertheless, in a recent work a systematic parametrization of the interaction in the frame of the pseudo-SU(3) model [27] for the $^{160-168}$Dy chain of dysprosium isotopes has allowed to calculate energies in the ground state, γ and β-bands. The calculations for these known nuclei were then extrapolated towards the neutron-rich nucleus ^{170}Dy.

Our starting points are the results obtained in a previous study [27]. In the present work, we use the eigenfunctions and eigenvalues of the Hamiltonian (2) of Ref. [27] in order to calculate the electric quadrupole transition strengths between states belonging to different bands and with values larger than 0.00001 e^2b^2. Our goal is to study the evolution of collectivity along the dysprosium isotopic chain (Z=66) starting at N = 94 and ending with the N = 104 midshell ^{170}Dy nucleus [28].

In contrast to the results reported [27] for the B(E2) strengths (between states belonging to the same band), in this contribution we are interested in the evaluation of cross transitions, that is between states belonging to different bands (inter-band). The calculation of these values is important in order to understand the nature of the excited 0^+ and 2^+ bands, which are usually refered as β and γ bands, respectively. As it has been pointed out by P. E. Garrett [29], the nomenclature used for the excited 0^+ bands should be done attending their properties, which involves the assessment of some B(E2) transitions between ground state, γ and β bands. In this way, the present calculations help to discriminate which of the 0^+ excitations in $^{160-170}$Dy correspond to β oscillations.

This contribution is organized as follows: The group theory that underlies the pseudo-SU(3) model is reviewed in the first part of the next section. Only a very limited number of results are given because a rather extensive discussion of the correspondence between the quantum numbers, Casimir operators, and pseudo-SU(3) wave functions can be found in previously published articles [20, 30]. The pseudo-SU(3) Hamiltonian and a choice for the corresponding interaction strength parameters are also discussed, as they are of special importance for a theoretical description of the $^{160-170}$Dy isotopes. In the final part of Sect. 2, definitions of experimentally measured quantities are reviewed. The pseudo-SU(3) model results for inter-band B(E2) transition strengths are presented in Sect. 3. A summary as well as a brief conclusion are given in Sect. 4.

2. The model

The pseudo-SU(3) model is a many-particle shell-model based theory that takes full advantage of pseudospin symmetry, which in heavy nuclei is manifest in the near degeneracy of the orbital pairs $[(l-1)_{j=l-1/2}, (l+1)_{j=l+1/2}]$. It is also a theory that takes full account of the Pauli Exclusion principle. Like most other shell-model schemes, the proton and neutron configuration spaces of the pseudo-SU(3) model are usually restricted to a single oscillator shell. A recent extension of the pseudo-SU(3) model takes the $\tilde{S} = 0$ and 1 (proton and neutron) spin degrees of freedom into account in a full proton-neutron formalism [25]. This extended theory was used to investigate the effect of different nuclear interactions on the low-energy properties of even-even and A-odd nuclei, where the influence of pairing and the remaining spin-orbit [22] correlations are known to play an important role.

The pseudo-SU(3) scheme is an algebraic shell-model theory that exploits powerful group theoretical methods in the construction of basis functions and for the calculation of required matrix elements. Specifically, basis states are labeled by eigenvalues of Casimir operators of the underlying symmetry groups and additional indices that are required to resolve multiplicities in the group reductions. Since a full discussion of these matters can be found elsewhere [7, 20], only the results will be included in this contribution. In terms of the space ($U(N) \leftrightarrow [f]$), shape ($SU(3) \leftrightarrow (\lambda, \mu)$), orbital ($SO(3) \leftrightarrow L$), spin ($S$) and total angular momentum (J) as well as the various multiplicities (α for the $U(N) \supset SU(3)$ reduction, ρ for the product of two SU(3) irreps, and κ for the $SU(3) \supset SO(3)$ reduction) the basis states have the form (subscript π for protons and ν for neutrons):

$$|\{\alpha_\pi(\lambda_\pi, \mu_\pi)\kappa_\pi S_\pi, \alpha_\nu(\lambda_\nu, \mu_\nu)\kappa_\nu S_\nu\}\rho(\lambda, \mu)\kappa L S; JM\rangle. \tag{1}$$

The occupation numbers for protons are constant along the chain. In shell model applications, the dysprosium isotopes are considered to have 16 protons out of the Z = 50 inert core, 10 of these in normal and 6 in abnormal $h_{11/2}$ parity levels. For neutrons, the occupations numbers in normal parity levels change from 8 to 14.

The Hamiltonian contains spherical Nilsson single-particle terms for the protons and neutrons ($H_{sp,\pi[\nu]}$), the quadrupole-quadrupole ($\tilde{Q} \cdot \tilde{Q}$) and pairing ($H_{pair,\pi[\nu]}$) interactions parametrized systematically, as well as three rotorlike terms (K^2, J^2 and C_3) that are diagonal in the SU(3) basis:

$$\begin{aligned} H &= \sum_{\alpha=\pi,\nu} \{H_{sp,\alpha} - G_\alpha \, H_{pair,\alpha}\} - \frac{1}{2}\,\chi\,\tilde{Q}\cdot\tilde{Q} \\ &\quad + a\,K^2 + \,b\,J^2 \,+\, c\,\tilde{C}_3. \end{aligned} \tag{2}$$

A detailed analysis of each term of this Hamiltonian and its parametrization can be found in Ref. [22]. The first row contains the basic components of any realistic Hamiltonian: the single-particle levels, pairing correlations and the quadrupole-quadrupole interaction, essential in the description of deformed nuclei. They have been widely studied in nuclear physics literature, allowing to fix their respective strengths by systematics. The rotorlike terms in Hamiltonian (2) are used to fine tune the spectra. Their three parameters a, b, and c were determined [27] by applying best fit to the experimental data for the five even-even isotopes $^{160-168}$Dy. The states from $J^\pi = 0^+$ to 6^+ belonging to ground, γ and β bands were used in the best fit procedure. In the present contribution, we are using the values a = 20, b = -3.2 and c = 0.033 (all in [keV]).

Before discussing comparisons of electromagnetic properties to experimental data, it is important to write down definitions for transition operators and other quantities.

The electric quadrupole transition operator, following the notation of Ref. [20], is defined as

$$\begin{aligned} T_{2\mu}(E2) &\equiv b_0^2 \left(e_\pi \sum_{i_\pi} \sqrt{\frac{16\pi}{5}} r_\pi^2(i) Y_{2\mu}(\theta_i^\pi, \phi_i^\pi) \right) \\ &+ b_0^2 \left(e_\nu \sum_{i_\nu} \sqrt{\frac{16\pi}{5}} r_\nu^2(i) Y_{2\mu}(\theta_i^\nu, \phi_i^\nu) \right). \end{aligned}$$

The SO(3) reduced matrix elements of a tensor operator T_{JM} between states of initial (final) angular momentum and projection J_i and M_i (J_f and M_f) is defined by

$$\langle \gamma_f J_f || T_J || \gamma_i J_i \rangle \langle J_i M_i, JM | J_f M_f \rangle \equiv \langle \gamma_f J_f M_f | T_{JM} | \gamma_i J_i M_i \rangle$$

where γ_i and γ_f represent additional quantum numbers that are required to uniquely define the initial and final states, and $\langle \gamma_f J_f || T_J || \gamma_i J_i \rangle$ stands for the reduced matrix element. It follows from this, that the reduced transition probability for electric quadrupole transition [30] is given by

$$B(E2; \gamma_i J_i \rightarrow \gamma_f J_f) \equiv \frac{2J_f + 1}{2J_f - 1} \langle \gamma_f J_f || Q_2 || \gamma_i J_i \rangle^2.$$

The definition of the electric quadrupole moment is given by the last expresion when $\gamma_i = \gamma_f$ and $J_i = J_f$. In that case, the electric quadrupole moment is given by

$$Q(\gamma J) \equiv \sqrt{\frac{16\pi}{5}} \sqrt{\frac{J(2J-1)}{(J+1)(2J+3)}} \langle \gamma J || Q_2 || \gamma J \rangle.$$

3. Inter-band B(E2) transition strengths

As indicated above, in this contribution the pseudo-SU(3) model is used to investigate electric quadrupole transition strengths in $^{160-170}$Dy and to compare with experimental data. The experimental energy spectra and their corresponding theoretical values have been published in [27].

Using the basis states, Hamiltonian and parametrization presented in Reference [27], we present the results for inter-band B(E2) strengths between states of ground state, γ and β-bands in the $^{160-170}$Dy isotopes in Table 1.

The effective charges used in the electric quadrupole transition operator Q_μ are e_π = 2.3 and e_ν = 1.3. These values are the same used in the pseudo-SU(3) studies up to now allowing to describe both intra- and inter-band B(E2)s. They are larger than those used in standard calculations of quadrupole transitions [28] due to the absence of nucleons in intruder levels, and they were not varied to fit any particular value. These results may be the starting point to realize a systematic analysis of the 0_2^+ excitations and to determine whether they are or not β

	B(E2) $[e^2b^2 \times 10^{-2}]$					
$J^{\pi}_{i,band} \rightarrow J^{\pi}_{f,band}$	^{160}Dy	^{162}Dy	^{164}Dy	^{166}Dy	^{168}Dy	^{170}Dy
$0^+_{g.s.b.} \rightarrow 2^+_{\gamma}$	14.401	14.603	16.501	2.306	1.703	18.401
$2^+_{g.s.b.} \rightarrow 4^+_{\gamma}$	1.577	1.602	0.672	0.346	0.358	0.507
$4^+_{g.s.b.} \rightarrow 5^+_{\gamma}$	4.034	4.070	3.369	0.041	0.037	4.734
$4^+_{g.s.b.} \rightarrow 6^+_{\gamma}$	0.190	0.186	0.081	0.105	0.102	0.081
$6^+_{g.s.b.} \rightarrow 7^+_{\gamma}$	2.133	2.166	1.081	*0.001	*0.001	2.596
$6^+_{g.s.b.} \rightarrow 8^+_{\gamma}$	0.021	0.153	1.227	0.037	0.013	0.259
$8^+_{g.s.b.} \rightarrow 9^+_{\gamma}$	0.643	0.659	0.121	0.048	0.050	1.333
$8^+_{g.s.b.} \rightarrow 10^+_{\gamma}$	0.021	0.912	2.479	0.042	0.144	1.039
$0^+_{g.s.b.} \rightarrow 2^+_{\beta}$	0.196	0.017	0.011	1.044	1.115	0.408
$2^+_{g.s.b.} \rightarrow 4^+_{\beta}$	0.313	*0.001	0.024	0.386	0.347	0.078
$4^+_{g.s.b.} \rightarrow 6^+_{\beta}$	0.068	0.003	0.063	0.024	0.027	0.035
$6^+_{g.s.b.} \rightarrow 8^+_{\beta}$	0.005	0.007	0.091	0.198	0.206	*0.001
$0^+_{\beta} \rightarrow 2^+_{\gamma}$	4.356	4.383	3.442	180.534	173.640	0.426
$2^+_{\beta} \rightarrow 3^+_{\gamma}$	1.374	1.080	1.775	108.443	113.550	19.116
$2^+_{\beta} \rightarrow 4^+_{\gamma}$	0.390	0.281	1.696	26.626	27.701	11.543
$4^+_{\beta} \rightarrow 5^+_{\gamma}$	0.317	0.320	1.968	1.092	0.996	26.994
$4^+_{\beta} \rightarrow 6^+_{\gamma}$	0.211	0.015	1.667	*0.001	0.014	33.639
$6^+_{\beta} \rightarrow 7^+_{\gamma}$	0.005	0.726	1.644	1.411	1.106	35.167
$6^+_{\beta} \rightarrow 8^+_{\gamma}$	0.043	0.002	1.640	1.139	0.374	57.099
$8^+_{\beta} \rightarrow 9^+_{\gamma}$	0.107	1.212	0.803	1.860	3.104	43.599

Table 1. Theoretical B(E2;$J_i^+ \rightarrow J_f^+$) [given in $e^2b^2 \times 10^{-2}$] inter-band transitions in $^{160-170}$Dy nuclei. Effective charges are e_{π}=2.3 and e_{ν}=1.3. The values with * are for B(E2;$J_i^+ \rightarrow J_f^+$) $<$ 0.001 $\times 10^{-2} e^2b^2$.

oscillations [29]. As it has been pointed out in this Reference, the $B(E2; 0^+_{gs} \rightarrow 2^+_{0^+_2})$ display a wide variation of values over well deformed nuclei, with variations of orders of magnitude over a single isotopic chain. That means that we can not continue our practice of labelling the 0^+_2 state by β, irrespective of its properties, because that practice has reduced the value of the term, and the original definitions of Bohr, Mottelson, Faessler and Greiner of the properties of a β vibration have been lost.

The calculation of the electric quadrupole moment and B(E2) transition strengths with realistic effective charges ($e_{\pi} = 1.4$ and $e_{\nu} = 0.6$) can be done if the intruder sector is explicitly considered. As discussed in Ref. [27], it can be obtained a simple evaluation of its effect by considering the quasi-SU(3) symmetry of the intruder sector, as proposed in Refs. [31, 32]. Having established for a given deformation the number of nucleons in the abnormal parity levels, their contribution to Q_0 is easily obtained by filling the orbits in the right panel of Fig. 34 in Ref. [33]. By summing this contribution and that of the normal parity sector, the B(E2) transitions are calculated with realistic values for effective charges e_{π} and e_{ν}. Nevertheless, that method does not allow to estimate the contribution of the intruder sector in each band; this means that the method only allows to calculate the constant contribution of the intruder sector for a given nucleus, and not allowing to differentiate its contribution in each band.

4. Summary and conclusions

For the first time, employing a systematically parametrized Hamiltonian and the best fit of three parameters for a set of nuclei [27], the model was used to study systematically inter-band B(E2) transitions in $160 < A < 168$ dysprosium isotopes and to predict the values between the ground, γ and β-bands in ^{170}Dy. The results show very collective bands and some degree of mixing between ground state and γ and β bands. This shows that the pseudo-SU(3) scheme is a very creditable theory for achieving a microscopic description of heavy deformed rare-earth nuclei. These results may be the starting point to realize a systematic analysis of the 0_2^+ excitations and to determine when they are or not β oscillations [29].

The results presented confirm the adequacy of the model to predict properties of rare-earth exotic nuclei. Experimental information for the neutron-rich nuclei is highly desired to understand nuclear structure around the midshell region. The ability of the model to describe complementary properties of heavy deformed nuclei such as g-factors, quadrupole moments and magnetic dipole transitions will be explored in the future.

4.1. Acknowledgments

This work was supported in part by CONACyT (México) under grant CB-2012-01-177519.

References

[1] Walker P M and Dracoulis G D 1999 *Nature (London)* **399** 35
[2] Wu C Y, Simon M W, Cline D, Davis G A, Macchiavelli A O and Vetter K 1998 *Phys. Rev.* C **57** 3466
[3] Firestone R B and Shirley V S 1996 *Table of isotopes* 8th edn (New York: Wiley)
[4] Bohr A and Mottelson B R 1998, *Nuclear Structure* (Nuclear Deformations vol 2) (Singapore: World Scientific Publishing Company)
[5] Casten R F and Zamfir N V 1993 *Phys. Rev. Lett.* **70** 402
[6] Zhao Y M, Arima A and Casten R F 2001 *Phys. Rev.* C **63** 067302
[7] Casten R F 1988 *J. Phys.* G: Nucl. Phys. **14**, S71
[8] M. Asai *et al.* 1999 *Phys. Rev.* C **59** 3060
[9] Jones E F *et al.* 2004 *J. Phys.* G: Nucl. Part. Phys. **30** L43
[10] Xiuqin L *et al.* 1998 *Eur. Phys. J.* A **2** 149
[11] Podolyak Z *et al.* 2000 *Proc. 2nd Intern. Conf. Fission and Properties of Neutron-Rich Nuclei* (St Andrews, Scotland) (Singapore: World Scientific) p 156
[12] Caamaño M 2002 *Ph.D. thesis* (U. of Surrey)
[13] Elliott J P 1958 *Proc. Roy. Soc. London Ser. A* **245** 128
[14] Elliott J P 1958 *Proc. Roy. Soc. London Ser. A* **245** 562
[15] Hecht K T and Adler A 1969 *Nucl. Phys.* A **137** 129
[16] Arima A, Harvey M and Shimizu K 1969 *Phys. Lett.* B **30** 517
[17] Ginocchio J N 1997 *Phys. Rev. Lett.* **78** 436
[18] Blokhin A L, Bahri C and Draayer J P 1995 *Phys. Rev. Lett.* **74** 4149
[19] Ratna Raju R D, Draayer J P and Hetch K T 1973 *Nucl. Phys.* A **202** 433
[20] Castaños O, Draayer J P and Leschber Y 1987 *Ann. Phys.* **180** 290
[21] Bahri C and Draayer J P 1994 *Comput. Phys. Commun.* **83** 59
[22] Vargas C E, Hirsch J G, Beuschel T and Draayer J P 2000 *Phys. Rev.* C **61** 031301(R); Vargas C E, Hirsch J G, and Draayer J P 2000 *Nucl. Phys.* A **673** 219
[23] Vargas C E, Hirsch J G and Draayer J P 2001 *Phys. Rev.* C **64** 034306
[24] Vargas C E, Hirsch J G and Draayer J P 2003 *Phys. Lett.* **B 551** 98
[25] Vargas C E and Hirsch J G 2004 *Phys. Rev.* C **70** 064320
[26] Hirsch J G, Popa G, Lesher S R, Aprahamian A, Vargas C E and Draayer J P 2006 *Rev. Mex. Fis.* S **52** 69
[27] Vargas C E, Velázquez V and Lerma S 2013 *Eur. Phys. J.* A **49** 4
[28] Ring P and Schuck P 1979 *The Nuclear Many-Body Problem* (Berlin: Springer)
[29] Garrett P E 2001 *J. Phys.* G: Nucl. Part. Phys. **27** R1
[30] Troltenier D, Bahri C and Draayer J P 1995 *Nucl. Phys.* A **586** 53; Troltenier D, Bahri C and Draayer J P 1995 *Nucl. Phys.* A **589** 75
[31] Zuker A P, Retamosa J, Poves A and Caurier E 1995 *Phys. Rev.* C **52** R1741
[32] Martínez-Pinedo G, Zuker A P, Poves A and Caurier E 1997 *Phys. Rev.* C **55** 187
[33] Caurier E, Martinez-Pinedo G, Nowacki F, Poves A and Zuker A P 2005 *Rev. Mod. Phys.* **77** 427

Journal of Physics: Conference Series **578** (2015) 012013 doi:10.1088/1742-6596/578/1/012013

Data-adaptive unfolding of nonergodic spectra: Two-Body Random Ensemble

R Fossion[1,2], G Torres Vargas[3], V Velázquez[4] and J C López Vieyra[5]

[1] Instituto Nacional de Geriatría, Periférico Sur No. 2767, 10200 México D.F., Mexico
[2] Centro de Ciencias de la Complejidad (C3), Universidad Nacional Autónoma de México, 04510 México, D.F., Mexico
[3] Posgrado en Ciencias Físicas, Universidad Nacional Autónoma de México, 04510 México, D.F., Mexico
[4] Facultad de Ciencias, Universidad Nacional Autónoma de México, 04510 México, D.F., Mexico
[5] Instituto de Ciencias Nucleares, Universidad Nacional Autónoma de México, 04510 México, D.F., Mexico

E-mail: fossion@nucleares.unam.mx

Abstract. The statistics of spectral fluctuations is sensitive to the unfolding procedure that separates global from local properties. Previously, we presented a parameter-free and data-adaptive unfolding method that we demonstrated to be highly effective for standard random-matrix ensembles from Random Matrix Theory (RMT). More general ensembles often break the ergodicity property, which leads to ambiguities between individual-spectrum averaged and ensemble-averaged fluctuation measures. Here, we apply our data-adaptive unfolding to a nonergodic Two-Body Random Ensemble (TBRE). In the present approach, both fluctuation measures can be calculated simultaneously within the same unfolding step, and possible arbitrarities introduced by traditional unfolding procedures are avoided.

1. Introduction

Standard matrix ensembles from Random Matrix Theory (RMT) [1], such as the Poisson ensemble and the Gaussian Orthogonal Ensemble (GOE), have been enormously successful in the modelling of the fluctuations of excitation spectra of quantum systems, that are regular and chaotic, respectively [2]. Recently, RMT has been applied as well in multivariate statistics to model the fluctuations of eigenspectra of correlation matrices of complex systems [3]. However, this RMT modelling is not completely realistic because many-body systems are effectively governed by one- and two-body forces, while canonical RMT implicitly assumes many-body forces between the constituents [4]. A stochastic modelling of the one- and two-body interaction yields a much smaller number of independent random variables, e.g., when a two-body random ensemble (TBRE) [5, 6], or a $k-$body embedded Gaussian ensembles (EGE) [7] is used. These more general ensembles introduce new statistical features that are absent in standard RMT, such as Gaussian instead of semicircular level densities [5] and nonergodicity [4, 5, 8, 9].

One of the difficulties in the statistical study of spectral fluctuations is the *unfolding* procedure which serves to separate the global level density density $\overline{\rho}(E)$ from the local fluctuations

Journal of Physics: Conference Series **578** (2015) 012013 doi:10.1088/1742-6596/578/1/012013

$\widetilde{\rho}(E) = \rho(E) - \overline{\rho}(E)$. The unfolding procedure is not trivial, and statistical results can be sensitive to the unfolding applied [10, 11]. In principle, the mean level density $\overline{\rho}$ can be determined as the average over all spectra of the ensemble, or – alternatively – as a running average over each spectrum individually. If the ensemble under study is *ergodic*, then the resulting *ensemble averaged* and *individual-spectrum averaged* fluctuation statistics are equivalent. The breaking of ergodicity creates an ambiguity in the characterization of the spectral fluctuations as the two measures lead to different results. In this case, the ensemble average is not representative for the individual realizations of the spectrum and an individual-spectrum unfolding is more appropriate [4, 5, 8]. When a correct unfolding is applied to a nonergodic ensemble that is chaotic, then spectral fluctuations of the GOE-type should result [9].

Previously, in Refs. [12, 13, 14], we proposed an unfolding method that is data-adaptive and model-free, and that expresses a spectrum in an exact way as the superposition of global and fluctuation normal modes. In this way, possible artifacts introduced by standard unfolding techniques are avoided. We presented two versions of the method. A first version is based on Singular Spectrum Analysis (SSA), which is a univariate time-series technique, and which can be applied to individual spectra to calculate individual-spectrum averaged statistics [12]. The SSA-based method includes one free parameter, but we demonstrated that the statistical results are parameter independent. The second version is based on a multivariate time-series technique, Singular Value Decomposition (SVD) [13], which is applied to an ensemble of spectra. The SVD-based method is preferrable as it does not include any parameters. Also, within this approach, both spectrum-averaged and ensemble-averaged fluctuation statistics can be calculated [14]. In Ref. [12], an SSA-based data-adaptive unfolding was succesfully applied to the individual spectra of a nonergodic TBRE ensemble. The purpose of the present contribution is to study an SVD-based data-adaptive unfolding of a TBRE ensemble.

2. SVD-based data-adaptive unfolding

Following the discussion of Ref. [13, 14], we consider an ensemble of $m = 1 \dots M$ excitation spectra $E^{(m)}(n)$, where each spectrum consists of $n = 1 \dots N$ levels. Each spectrum can be accomodated in a row of the $M \times N$ dimensional matrix $\mathbf{X}$, which is interpreted as a multivariate time series,

$$\mathbf{X} = \begin{pmatrix} E^{(1)}(1) & E^{(1)}(2) & \cdots & E^{(1)}(N) \\ E^{(2)}(1) & E^{(2)}(2) & \cdots & E^{(2)}(N) \\ \vdots & \vdots & \ddots & \vdots \\ E^{(M)}(1) & E^{(M)}(2) & \cdots & E^{(M)}(N) \end{pmatrix}. \tag{1}$$

Singular Value Decomposition (SVD) decomposes $\mathbf{X}$ in a *unique* and *exact* way as,

$$\mathbf{X} = \mathbf{U}\boldsymbol{\Sigma}\mathbf{V}^T = \sum_{k=1}^{r} \sigma_k \mathbf{X}_k = \sum_{k=1}^{r} \sigma_k \vec{u}_k \vec{v}_k^T, \tag{2}$$

where $\boldsymbol{\Sigma}$ is an $M \times N$-dimensional matrix with only diagonal elements which are ordered *singular values* or weights $\sigma_1 \geq \sigma_2 \geq \ldots \geq \sigma_r$, with $r \leq \text{Min}[M, N] = \text{rank}(\mathbf{X})$. The vectors $\vec{v}_k$ are orthonormal, and correspond to the kth columns of the $N \times N$-dimensional matrix $\mathbf{V}$. They are *normal modes*, and they constitute a basis for the row space of $\mathbf{X}$. The vectors $\vec{u}_k$ are orthonormal as well, corresponding to the kth columns of the $M \times M$-dimensional matrix $\mathbf{U}$, and can be interpreted as the associated *projection coefficients*. The elementary matrices $\mathbf{X}_k = \vec{u}_k \vec{v}_k^T \equiv \vec{u}_k \otimes \vec{v}_k$ are calculated from the outer product of $\vec{u}_k$ and $\vec{v}_k$. In this way, any matrix row of **X** containing a particular excitation spectrum $E^{(m)}(n)$ can be written as,

$$E^{(m)}(n) = \sum_{k=1}^{r} \sigma_k U_{mk} \vec{v}_k^T(n). \tag{3}$$

The square of a singular value or weight gives a *partial variance*, $\lambda_k = \sigma_k^2$, which is the fraction of the total variance $\lambda_{\text{tot}} = \sum_{k=1}^{r} \lambda_k$ of the multivariate time series **X** carried by the normal mode $\vec{v}_k$.

A first result of Ref. [13] is that the main feature of a typical energy sequence $E(n)$ is a dominant monotonous trend $\overline{E}(n)$, whereas the superposed fluctuations $\widetilde{E}(n) = E(n) - \overline{E}(n)$ are usually orders of magnitude smaller. Consequently, the global properties of a spectrum are represented by non-oscillating normal modes $\vec{v}_k$ associated to the first few $k = 1, \ldots, n_T$ dominant partial variances λ_k in the so-called *scree diagram* of ordered partial variances, which can be distinguished easily from the much smaller higher-order partial variances λ_k with $k = n_T + 1, \ldots, r$ corresponding to oscillating normal modes $\vec{v}_k$ associated to the fluctuations.

Another result from Ref. [13] is that when the data-adaptive unfolding of Eq. (2) is applied to the standard Poisson or GOE matrix ensembles, the part of the scree diagram associated to the fluctuation normal modes obeys the power law,

$$\lambda_k \propto 1/k^{\gamma}, \tag{4}$$

indicating that the long-range correlations of the fluctuations are scale invariant (fractal). The exponent γ characterizes the *rigidity* of the fluctuations, with $\gamma = 2$ in the case of soft spectra (Poisson), and $\gamma = 1$ in the case of rigid spectra (GOE). On the other hand, any particular spectrum from the ensemble can be unfolded separately using Eq. (3), after which its individual fluctuations can be studied with, e.g., Fourier spectral analysis. In the case of the standard random-matrix ensembles, the Fourier power spectrum of the fluctuations of the individual eigenspectra also obeys a power law,

$$P(f) \propto 1/f^{\beta}, \tag{5}$$

with $\beta = \gamma = 2$ (Poisson) and $\beta = \gamma = 1$ (GOE), such that the scale invariance property does not seem to depend on the particular basis used for the decomposition. In Ref. [14], it was argued that Eq. (4) can be interpreted as an ensemble-averaged fluctuation measure, because it quantifies the scaling properties of the normal modes in common to all the spectra of the ensemble, whereas Eq. (5) was interpreted as an individual-spectrum averaged fluctuation measure. Note that within the present SVD-based approach, both measures can be calculated within one and the same unfolding procedure. In contrast, in the traditional approach, each measure is obtained after a separate unfolding procedure which makes a consistent comparison between both measures difficult (see, e.g., [4]).

3. Two-Body Random Ensemble (TBRE)

In the present contribution, we consider an ensemble of theoretical excitation spectra for the ^{48}Ca atomic nucleus. The spectra are calculated with a nuclear shell-model code, based on ANTOINE [20], and using random interactions (TBRE [5, 21]). The nucleus ^{48}Ca is an obvious choice to study within the nuclear-shell model framework, because an inert ^{40}Ca core can be assumed, and calculations can be carried out in the full neutron fp shell. The level fluctuations of ^{48}Ca have been thoroughly studied in literature and have been found to correspond with GOE

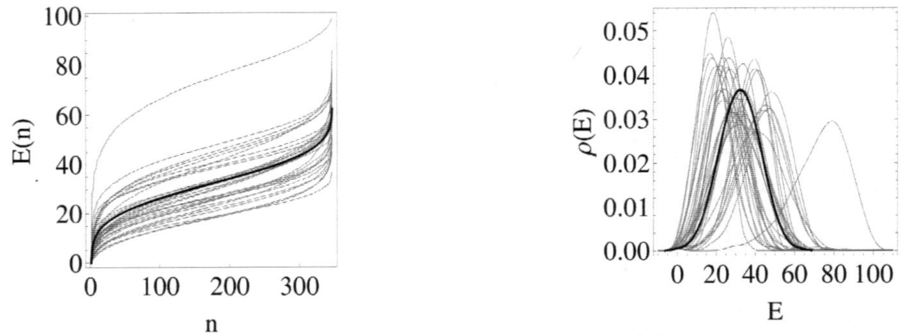

(a) Energy sequence (b) Level density

Figure 1. TBRE ensemble of $M = 30$ spectra, each with $N = 347$ levels. (a) Energy sequences $E^{(m)}(n)$ with $n = 1, \ldots, N$ and $m = 1, \ldots, M$. (b) Level densities $\rho(E)$. It can be appreciated that the ensemble mean (thick black curve) is not representative for the individual realizations of the spectrum (gray curves).

predictions [15, 16, 17, 18, 19], which permits to verify the validity of the present SVD unfolding method. Fluctuation studies must be applied to levels with the same quantum numbers; we arbitrarily choose the $J^\pi = 0^+$ subspectrum. The ensemble consists of $m = 1, \ldots, M$ energy sequences $\{E^{(m)}(n), n = 1, \ldots, N\}$ with $N = 347$ and $M = 30$, see Fig. 1. Shown are the energy sequences $E(n)$ (panel a) and also the level densities $\rho(E)$ (panel b). Note that the level densities have a gaussian shape instead of obeying the semicircle distribution typical for GOE ensembles. It can be seen that the ensemble average is not representative for the individual realizations of the excitation spectrum, which is an indication of the nonergodicity property of this ensemble [4, 5]. To take into account only the bulk (central part) of the spectrum when performing the SVD-based unfolding procedure, 10% of the lower and upper levels will be discarded. In Ref. [12], when the SSA-based data-adaptive unfolding was applied to the individual spectra of this TBRE ensemble, a scree diagram and Fourier power spectrum were obtained that obey the power laws of Eq. (4) and Eq. (5) with exponents $\gamma \approx \beta \approx 1$, indicating the expected GOE-type fluctuations, in confirmation with results from other authors [5, 21].

4. Results

Applying the SVD-based data-adaptive unfolding of Eq. (2) to the TBRE ensemble of Fig. 1, a scree diagram of ordered partial variances λ_k is obtained, where λ_k with $k = 1, \ldots, 3$ are orders of magnitude larger than the higher-order λ_k, see Fig. 2(a). Inspection of the associated normal modes $\vec{v}_k$, confirms that $\vec{v}_k$ with $k = 1, \ldots, 3$ do not oscillate, whereas the higher-order normal modes $\vec{v}_k$ do oscillate, see Fig. 2(b). The former suggests that the global spectral properties of the TBRE ensemble are described by the first $n_T = 3$ normal modes. The higher-order partial variances λ_k with $k = n_T + 1, \ldots, r$ follow the power law of Eq. (4) with $\gamma \approx 1.83$, which is much closer to the Poisson limit than to the GOE limit, and which is in contradiction with the result $\gamma \approx 1$ from Ref. [12]. This result seems to confirm that, in the case of nonergodic ensembles, the ensemble-averaged fluctuation measure of the power law of Eq. (4) of the scree diagram is not reliable. The close-to Poissonian statistics appears to indicate that the normal modes $\vec{v}_k$ are rather uncorrelated. To check this idea, we consider the participation ratio PR(k),

$$\mathrm{PR}(k) = 1/\sum_{m=1}^{M} U_{km}, \tag{6}$$

Journal of Physics: Conference Series **578** (2015) 012013 doi:10.1088/1742-6596/578/1/012013

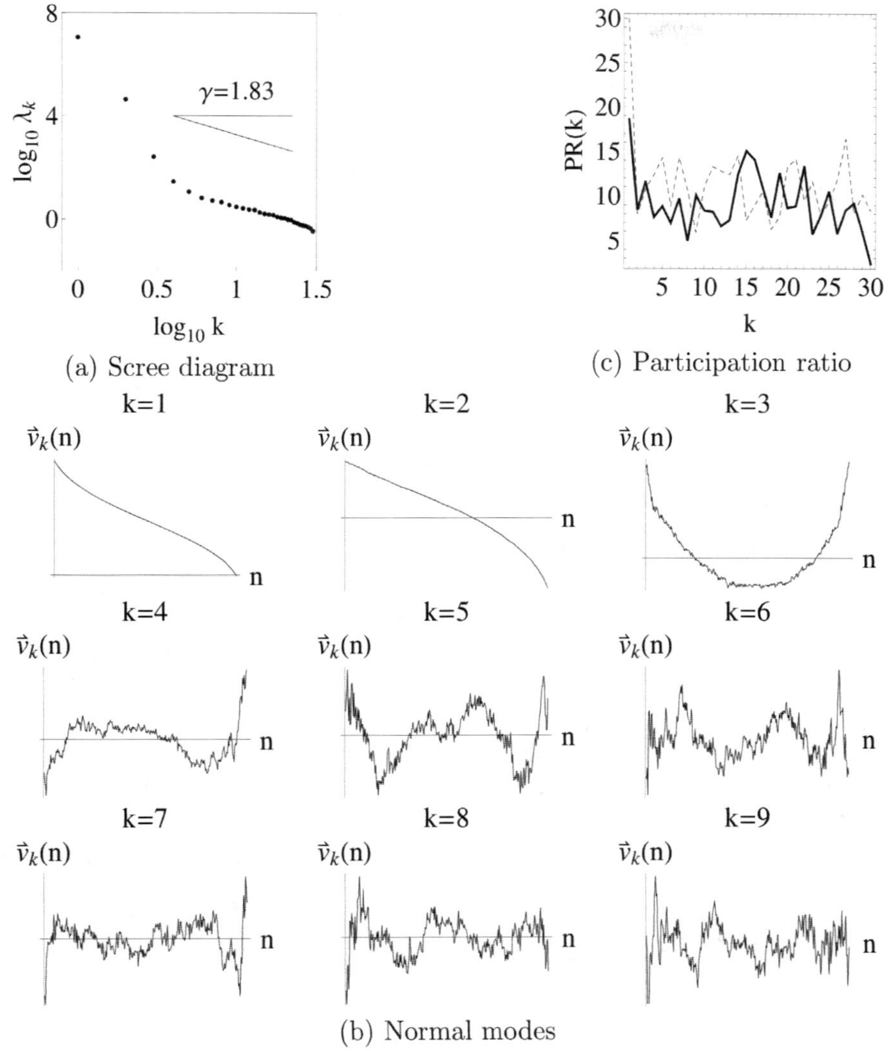

Figure 2. SVD decomposition of the ensemble of TBRE spectra of Fig. 1. (a) Scree diagram of ordered partial variances λ_k, where the λ_k with $k = 1, \ldots, n_T$ describes the global spectral properties, the λ_k with $k = n_T + 1, \ldots r$ obey the power law of Eq. (4) with $\gamma = 1.83$ and describes the local spectral properties. (b) First 9 normal modes $\vec{v}_k$. (c) Participation ratio PR(k) of the normal modes $\vec{v}_k$ of the TBRE ensemble (thick curve) compared to an equivalent GOE ensemble (dashed curve).

which gives the number of spectra of the TBRE ensemble to which the normal mode $\vec{v}_k$ significantly contributes [22]. In Fig. 2(c), the participation ratio PR(k) of the TBRE ensemble is compared with the participation ratio of a GOE ensemble with the same dimensions M and N. Overall, it can be appreciated that the normal modes $\vec{v}_k$ contribute to a larger number of spectra in the ergodic GOE case than in the nonergodic TBRE case. It can be appreciated, however, that the most important difference resides in the first dominant normal mode $\vec{v}_1$, which contributes to all $M = 30$ spectra in the GOE case, and to only two thirds of the spectra in the TBRE case. As each specific normal mode does contribute to some spectra and does not contribute to some other spectra, the set of normal modes partially decorrelates, and their scaling properties deviates towards Poisson statistics.

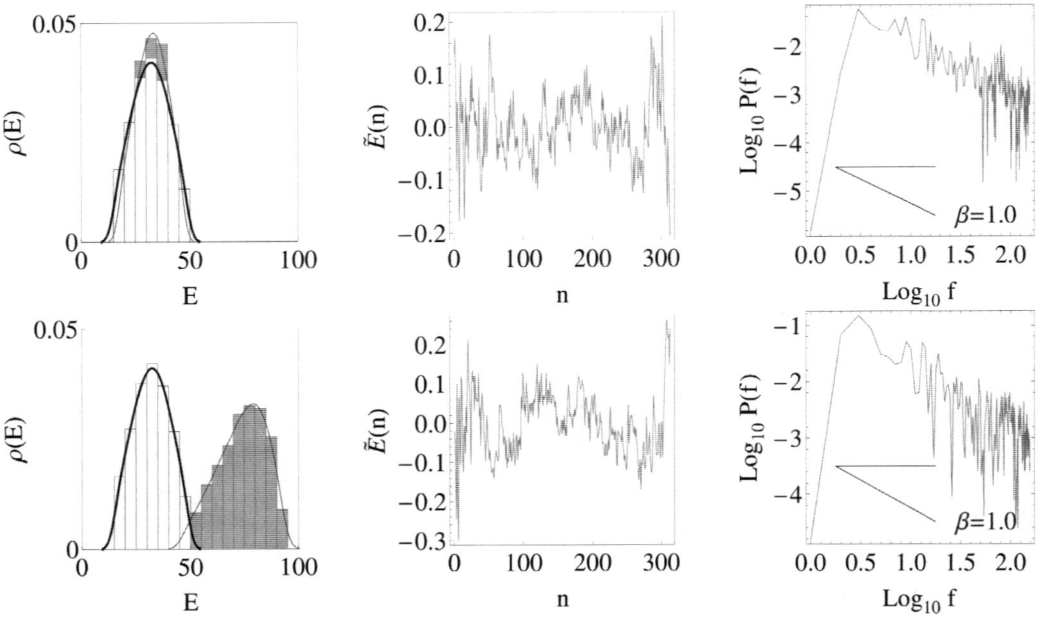

(a) Level density (b) Fluctuations (c) Power spectrum

Figure 3. Data-adaptive unfolding of the TBRE ensemble of Fig. 1. Results of the unfolding are shown for two different spectra; one spectrum lies close to the ensemble average (upper row) and the other spectrum lies far from the ensemble average (lower row). (a) Empirical level density of the individual spectrum (gray histogram) compared to the ensemble average (white histogram); compared to the data-adaptive global level density $\rho(\overline{E})$ for the realization (thin curve) and for the ensemble average (thick curve). (b) Fluctuations of the realization. (c) The Fourier power spectrum of the fluctuations of the realization behaves as the power law of Eq. (5) with $\beta = 1$.

Using the expression of Eq. (3), each separate spectrum can be expressed exactly as a weighted sum of normal modes, such that also an individual-spectrum averaged fluctuation measures can be calculated. The global part of an individual spectrum is given by [13],

$$\overline{E}^{(m)}(n) = \sum_{k=1}^{n_T} \sigma_k U_{mk} \vec{v}_k(n), \tag{7}$$

and the local fluctuating part can be expressed as [13],

$$\widetilde{E}^{(m)}(n) = \sum_{k=n_T+1}^{r} \sigma_k U_{mk} \vec{v}_k(n). \tag{8}$$

In Fig. 3, two specific spectra from the ensemble are represented. The first spectrum is selected from a region close to the ensemble average, and the second spectrum is chosen far from the ensemble average. Shown are (a) the histogram of the levels of the spectrum compared to the SVD-based data-adaptive global level density $\rho(\overline{E})$, (b) the local fluctuations $\widetilde{E}(n)$, and (c) the corresponding Fourier power spectrum $P(f)$. It can be appreciated that the Fourier power spectrum of the fluctuations of the individual spectra obeys the power law of Eq. (5) with exponent $\beta \approx 1$, in correspondence with the results with SSA-based data-adaptive unfolding [12], and in correspondence with results from other authors using traditional unfolding [5, 21].

5. Conclusion

In the present contribution, we applied a SVD-based data-adaptive unfolding technique that we first presented for standard ergodic random-matrix ensembles from RMT to a nonergodic TBRE ensemble. Ensemble averaged and individual-spectrum averaged statistics can be calculated in a parameter-free and consistent way within the same unfolding procedure. In a forthcoming publication, we will apply the SVD-based unfolding to a random-matrix model with tunable nonergodicity to study its effect on the statistical properties [23].

Acknowledgements

The authors acknowledge financial support from CONACYT (Grants No. CB-2011-01-167441). This work was also partly funded by the Instituto Nacional de Geriatría (project DI-PI-002/2012). The authors wish to thank also the anonymous referee of our previous publication [13] who brought the interesting topic of nonergodicity to our attention.

References

[1] Mehta M L 1991 *Random matrices* (Acad. Press, New York, 2nd ed)
[2] Haake F 2010 *Quantum signatures of chaos* (Springer, Heidelberg, 3rd ed)
[3] Kwapień J and Drożdż S 2012 *Phys. Rep.* **515** 115
[4] Asaga T, Benet L, Rupp T and Weidenmüller H A 2001 *Europhys. Lett.* **56** 340
[5] Flores J, Horoi M, Müller M and Seligman T H 2001 *Phys. Rev. E* **63** 026204
[6] French J B and Wong S S M 1970 *Phys. Lett. B* **33** 449; Bohigas O and Flores J 1971 *Phys. Lett. B* **34** 261; French J B and Wong S S M 1971 *Phys. Lett. B* **35** 383
[7] Mon K K and French J B 1975 *Ann. Phys.* **95** 90; Benet L and Weidenmüller H A 2003 *J. Phys. A: Math. Gen.* **36** 3569
[8] Jackson A D, Mejia-Monasterio C, Rupp T, Saltzer M and Wilke T 2001 *Nucl. Phys. A* **687** 405
[9] Bohigas O, de Carvalho J X and Pato M P 2008 *Phys. Rev. E* **77** 011122
[10] Gómez J M G, Molina R A, Relaño A and Retamosa J 2002 *Phys. Rev. E* **66** 036209
[11] Abuelenin S M and Abul-Magd A Y 2012 *Procedia CS* **12** 69
[12] Torres Vargas G, Fossion R, Velázquez V and López Vieyra J C 2014 *J. Phys.: Conf. Ser.* **492** 012011
[13] Fossion R, Torres Vargas G and López Vieyra J C 2013 *Phys. Rev. E* **88** 060902(R)
[14] Fossion R, Torres Vargas G and López Vieyra J C 2013 submitted to *Phys. Rev. E*, arXiv:1402.6786 [nlin.CD]
[15] Gómez J M G, Kar K, Kota V B K, Molina R A, Relaño A and Retamosa J 2011 *Phys. Rep.* **499** 103
[16] Landa E, Morales I O, Hernández C, López Vieyra J C and Frank A 2008 *Rev. Mex. Fís.* S **54** 48
[17] Landa E, Morales I O, Fossion R et al. 2011 *Phys. Rev. E* **84** 016224
[18] Gómez J M G et al. 1998 *Phys. Rev. C* **58** 2108.
[19] Caurier E , J.M.G. Gómez, V.R. Manfredi and L. Salasnich et al. 1996 *Phys. Lett. B* **365** 7
[20] Caurier E and Nowacki F 1989 code Antoine, Strasbourg
[21] Relaño A, Molina R A and Retamosa J 2004 *Phys. Rev. E* **70** 017201
[22] Plerou V, Gopikrishnan, Nunes Amaral L A and Stanley H E 1999 *Phys. Rev. Lett.* **83** 1471; Plerou V, Gopikrishnan P, Rosenow B, Nunes Amaral L A, Guhr T and Stanley H E 2002 *Phys. Rev. E* **65** 066126
[23] Fossion R, in preparation, arXiv:1401.0867 [nlin.CD]

XXXVII Symposium on Nuclear Physics IOP Publishing
Journal of Physics: Conference Series **578** (2015) 012014 doi:10.1088/1742-6596/578/1/012014

Quantum interference vs. quantum chaos in the nuclear shell model

Gerardo Fernández[1], Edna M Hernández[2], M Hautefeuille[1], E Landa[3,4], I O Morales[3,4], V Velázquez[1], R Fossion[4,5], C E Vargas[6] and A Frank[3,4]

[1]Facultad de Ciencias, Universidad Nacional Autónoma de México, Apartado Postal 70-543, 04510 México, D.F., México
[2] Laboratoire de Chimie de Coordination, CNRS and Université de Toulouse (UPS, INP), 31077 Toulouse, France
[3] Instituto de Ciencias Nucleares, Universidad Nacional Autńoma de México, Apartado Postal 70-543, 04510 México, D.F., México
[4] Centro de Ciencias de la Complejidad (C3), Universidad Nacional Autónoma de México, 04510 México, D.F., México
[5] Instituto Nacional de Geriatría, Periférico Sur No. 2767, 10200 México D.F., México
[6] Facultad de Física e Inteligencia Artificial, Universidad Veracruzana. Lomas del Estadio S/N; Zona UV, Xalapa, Ver., 91000, México

E-mail: `vicvela@ciencias.unam.mx`

Abstract. In this paper we study the complexity of the nuclear states in terms of a two body quadupole-quadrupole interaction. Energy distributions and eigenvectors composition exhibit a visible interference pattern which is dependent on the intensity of the interaction. In analogy with optics, the visibility of the interference is related to the purity of the states, therefore, we show that the fluctuations associated with quantum chaos have as their origin the remaining quantum coherence with a visibility magnitude close to 5%.

1. Introduction

Starting with the premise that every classical mechanics state (C) may be described by a subjacent theory based on quantum mechanics (Q), the transition from quantum to classical behavior and vice-versa deserves to be studied; see [1] for example. What are the characteristics of such a transition? What is lost of gained in this process? Is the transition a sudden shift between two phases or does it take place in a continuous manner? There are questions that are often raised regarding the bridge joining classical to quantum worlds. First of all, we have some clear ideas about how those systems behave in corresponding phases. We know that if a quantum system has access to a set of conditions its state is a linear superposition of all of those, for instance. This is known as the spectral decomposition principle [2]. Therefore, it is said that a physical state coexists in all the states of the base in which it has been described with a certain amplitudes of probability distribution. If this amplitudes of probability distribution is fixed, the state is considered a quantum one; if it is random (or mixed statistics) it is considered classical. This quantum coexistence of states permits that some phenomena such as auto-interference and quantum entanglement are properties without any classical analogies. In the $Q \to C$ transition,

Journal of Physics: Conference Series **578** (2015) 012014 doi:10.1088/1742-6596/578/1/012014

the quantum state is converted into a classical state when coexistence cannot be traced back (this does not mean that it does not exist). According to Zurek [1] and other authors argument [3, 4, 5] this transition finds its origin in the interaction of the quantum ensemble with the surrounding environment. This phenomenon is known as decoherence. Decoherence is thus caused by the information exchange between the system (S) and its surrounding environment (E). When the environment is classical, the exchange will not be controlled and the quantum (fixed) amplitudes of probability distribution will be thermal, turning to random one, thus classical. In this paper, we show that this information exchange is taking place via an interaction. Every interaction process implies mixed quantum components. In the nuclear shell model framework, we suppose that each nucleon interact with a quantum environment, then, information is developed through two mechanisms: the number of times the nucleon interacts with the environment in an average time, and the interaction strength. We will use concepts such as quantum interference as a property of the states purity and show how this property is lost with the mixing of configurations. As a conclusion of this analysis we propose that quantum fluctuations of the nuclear energy levels, widely studied in the framework of quantum chaos (QC), find their origin in the residuum of quantum coherence, nearly totally lost due to multiple interactions between the states of the configuration space. Because we consider a quantum isolated environment our transition will be $Q \rightarrow QC$ For this goal we develop shell model calculations using a tunable two body quadrupole-quadrupole interaction.

2. The Model

The most important ingredient of any model is its interaction. In the nuclear shell model, the interaction is modeled as the sum of an mean interaction and an interaction of N bodies, respectively called monopolar (m) and multipolar (M).

$$\hat{H}_{SM} = \hat{H}_m + \hat{H}_M \tag{1}$$

The simplest representation of this sum is the field of an independent particle (monopolar part) plus the interaction of two bodies (multipolar part). In practice, the diagonal part of the interaction of two bodies is usually included in the monopolar part. We have selected a schematic interaction such that we can observe the influence of the mixed states on the $Q \rightarrow QC$ transition via the quadrupolar interaction parameter. We thus considered an interaction of two bodies composed of quadrupolar and pairing interactions. The monopolar interaction will be composed of the mean field of a harmonic oscillator including the diagonal part of the modified realistic Kuo and Brown interaction, KB3 [6].

$$\hat{H}_{SM} = \hat{H}_m - \chi\hat{Q}\cdot\hat{Q} + g\hat{P}\cdot\hat{P} \tag{2}$$

The pairing interaction coupling constant value will be set to $g = 0.46$ [7] and the quadrupolar interaction parameter will be swept within an acceptable interval. This interaction will then be applied to the ^{48}Ca nucleus problem in the fp shell. We will use the shell model code ANTOINE [8] with a Lanczos diagonalization process. In particular, we will calculate the statistics of 1627 states with $J^p = 3^+$. Figure 1 presents the energy sequences for each diagonalization of the Hamiltonian (2) with a dependency on the quadrupolar interaction intensity parameter χ. It can be seen that for lower values of χ the sequence is discontinued; it presents energy jumps. As the interaction of two bodies increases, the empty spaces fill themselves and the energy series curves are flattened. Figure 2 shows the distribution of energy levels for four quadrupolar force intensities: $\chi = 0.01$; (b) $\chi = 0.04$; (c) $\chi = 0.07$ y (d) $\chi = 0.25$.The distribution of energies present peaks that disappear as the interaction increases, as can be seen on these graphs. The peaks represent the energy leaps mentioned in the first series of figure 1 and are energy ensembles that have not been mixed with the energies of other peaks. The quadrupolar interaction is a long-distance interaction, but its range is limited at very low intensities. Therefore, each peak is a local distribution of energy in which interactions appear between energies of the sub-ensemble [11].

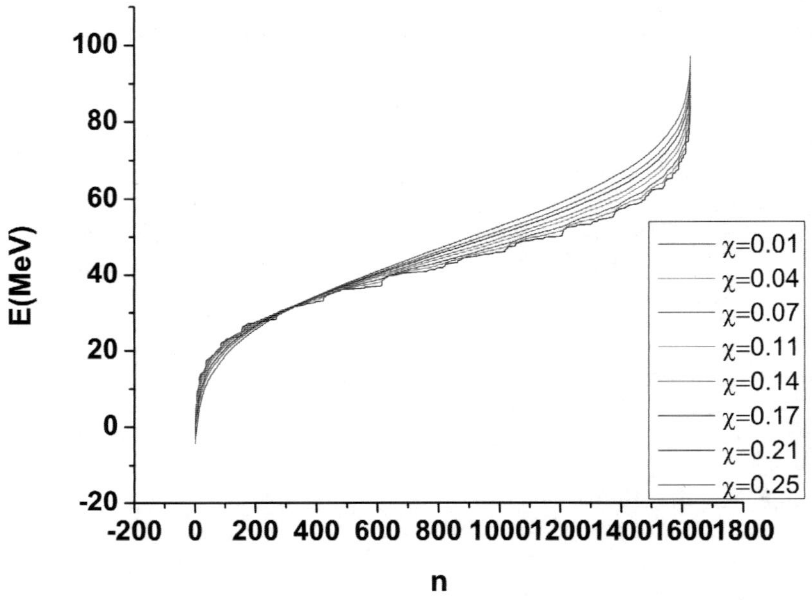

Figure 1. Energies of the 1627 states with $J = 3+$ in ^{48}Ca with dependence of the quadrupole-quadrupole parameter χ. For weak interactions is possible to observe jumps in the sequences of energies.

3. Energy interference

The 1627 states are constrained to certain spectral zones, which allows us to say that the spectrum present some form of organization. Energies are thus organized in sub-ensembles that share the common property that at low interaction intensities their particular states are not mixed with the states of another sub-ensemble. This can be interpreted as interference. In fact, the distribution of energies forms an interference pattern. The definition of interference indeed implies that it exists if the states that produce it are identical or at least present mutual coherence. The difference between the four graphs of figure 2*a* lies in their respective degree of coherence: it is decreases from figure 2*a* to figure 2*d*.

Technically, the nucleus state in the sub-space $J^p = 3^+$ can be written in a certain base $|k>$:

$$|\psi^i >= \sum_{k=1}^{1627} C_k^i |k> \tag{3}$$

where each one of the energies E_i (i from 1 to 1627) E_i is a solution of:

$$\hat{H}|\psi^i >= \bar{E^i}|\psi^i > \tag{4}$$

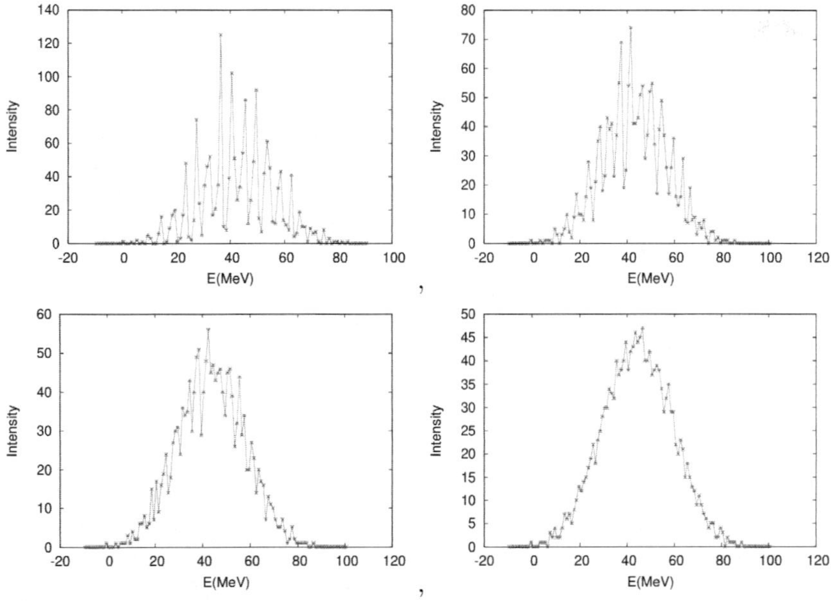

Figure 2. Energy distribution plots for four of the eight energy series showed in the Fig 1. : a) $\chi = 0.01$, b) $\chi = 0.04$, c) $\chi = 0.07$, y d) $\chi = 0.25$

if the states are not mixed statistically with each other. In this case we say the states are pure: $\bar{E^i} = E^i$ If the states are not pure, the energy of each state will be the mean of of the states energies E^i that compose the mixed statistics. The deviation around the mean is determined by by the coefficients C^j_k. The maximum degree of purity that each state can present is $|C^i_k|^2 = \delta_{i,k}$ In principle, the mean field already broke the degeneration of the harmonic oscillator field; however, it is possible to say that the states maintain a certain degree or purity as gaps are observed in the energy density.

4. Power spectrum and visibility

As the quadrupolar interaction intensity increases the states mixing increases and the gaps gradually disappear. When the interaction increases up to one value similar to the force of the realistic interaction, the interference is reduced down to fluctuations (mainly visible around the maximum of the distribution, figure 2*d*.). The meaning of these fluctuations as the remainder of a certain coherence can be interpreted back (from distribution to spectrum) as the fluctuation of each energy around its mean. How can such coherence be measured? In an effort to make an optical analogy [9], the difference between a coherent and a non-coherent state can be expressed as a spectral width. Such spectral width is inversely proportional to the coherence length and greater coherence guarantees greater interference and greater state purity. In the nuclear model, the coherence level of states follows a similar rule: lower configuration mixing (or greater purity) implies greater coherence. To follow this optical analogy, it is possible to define the visibility of the energy distribution as:

$$V = \frac{I_{max} - I_{min}}{I_{max} + I_{min}} \tag{5}$$

where I_{max} and I_{min} are referring to the intensities (or counts numbers) from a peak to a consecutive valley in a distribution.

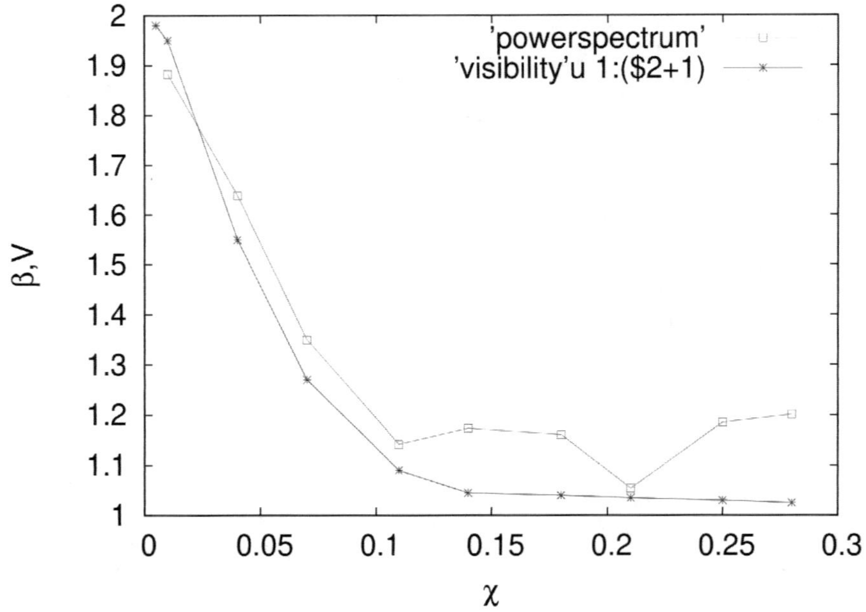

Figure 3. Visibility and power spectra parameter for energy time series dependent of the quadrupole-quadrupole parameter χ. In order to compare, visibility was increased by one.

Figure 3 shows both the visibility (asterisks) and power spectrum parameter (squares) as functions of the quadrupolar interaction. As visibility is related to state purity, it is possible to say that state mixing increases exponentially and that half of the total mixing is attained for a quadrupolar interaction of $\chi = 0.045$. Other interesting aspect is that for a standard quadrupolar interaction ($\chi = 0.21$) the visibility is around 5%. We can then ask ourselves if a visibility of 5% only represents noise or if there still is some degree of correlation. In order to investigate this point, we compare with the power spectra fluctuation analisys. By subtracting the mean behavior (in all curves in Fig. 1) we can obtain the energy time series of the fluctuations. In [10, 11] the method of power spectrum analysis (PS) [14, 15, 16, 17] is used to study the statistical of fluctuations of energy time series. Following the $1/f^\beta$ noise convention, three of the most conventional noises present in nature are for $\beta = 2, 1, 0$ corresponding to thermal (Brownian), chaotic and Poisson noise respectively [12]. PS curve shows the behavior of the noise analyzed with the power spectrum method as a function of the quadrupolar interaction parameter: β vs. χ. For low values of χ, the noise appears to be $1/f^2$, and it evolves to a $1/f$ noise as interaction increases. There is a clear similarity between noise behavior and visibility: both show an exponential regime in the same $0, 0.1$ χ interval.

It is also interesting to remark that the statistics of the time series is shifted to a $1/f$ noise, corresponding to approximately a visibility of 5%. It is thus tempting to conjecture that a zero visibility would correspond to a $1/f^0$ noise, or a random noise. Although in reference [13] may be argued otherwise, we never find any random noise.
With the convention that 100% visibility corresponds to totally pure states with a maximal coherence length (all the states of the ensemble correspond to one single state ψ^i of the base), zero visibility would correspond to the states of an ensemble with a null coherence length represented by a random mixing of all the states of the base. Hence, time series with interactions such that $\chi > 0.1$, still possess a certain degree of coherence. In this case, the $1/f$ noise, characteristic of

Journal of Physics: Conference Series **578** (2015) 012014 doi:10.1088/1742-6596/578/1/012014

nuclear spectrum [14, 15, 16, 17, 10], corresponds to the same fluctuations found in the analysis of the eigenvalues of random matrices [18, 19, 20] of a two-body random ensemble (TBRE). Studies of these random matrices showed that nuclear energy levels of low-energy excited states of complex nuclei follow a complete statistics, and not a random one. This complexity is called quantum chaos. We can propose that such fluctuations still possess information, given the correspondence. This information is stored in the remaining coherence that originates the 5% visibility.

5. Decoherence and conclusions

To analyze this remaining coherence, the dependence of eigenvectors with the quadrupolar interaction parameter has been studied. Figures 4 and 5 show the distribution of the squares of the 1627 quantum components ($|C_K^\alpha|^2$) of the 1627 states $|\psi_K>$ with $J^p = 3^+$ in ^{48}Ca for $\chi = 0.01$ and $\chi = 0.21$ respectively. It can be observed that in the first case (fig.4) the components distribution is fractioned, the states are still organized in groups. As previously discussed, the states are locally mixed for this level of interaction. In other words, we can say that components of a given states group have not mixed with components of another group. However, as interaction increases the groups vanish showing a more homogeneous mixing (fig.5). It is worth mentioning that the envelope of such a distribution presents similarities with the information distribution obtained from the entropy $S(\alpha) = -\sum_K |C_K^\alpha|^2 log|C_K^\alpha|^2$ where ψ^α is the state described in the $|K>$ base states [21], [22]. The shape of this global distribution tells us that the states furthest from the energy mean are very close to what could be considered as pure states: only very few components are present in the superposition. The particular case of the 3^+ state of less energy possess around ten components, corresponding to a degree of purity (seen geometrically) of approximately 94% ($[1-(10/1627)]100$). As the state-mixing increases, the number of components involved in the description of the state increases. The principal components form a group with a maximum number in the central states.

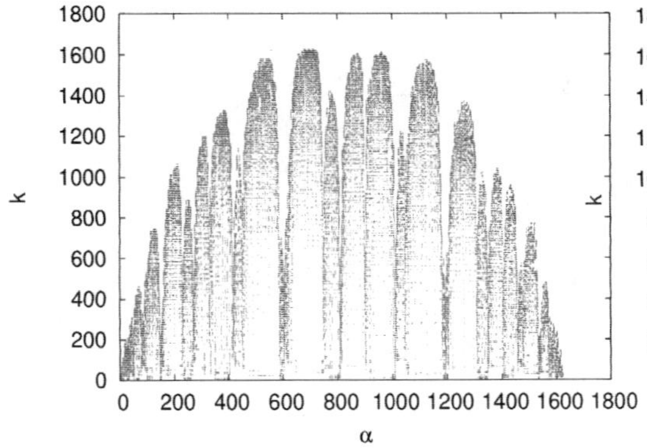

Figure 4. Distribution of probabilities $|C_k^\alpha|^2$ for all 1627 α states. For $\chi = 0.005$ it is appreciable the interference among basis states.

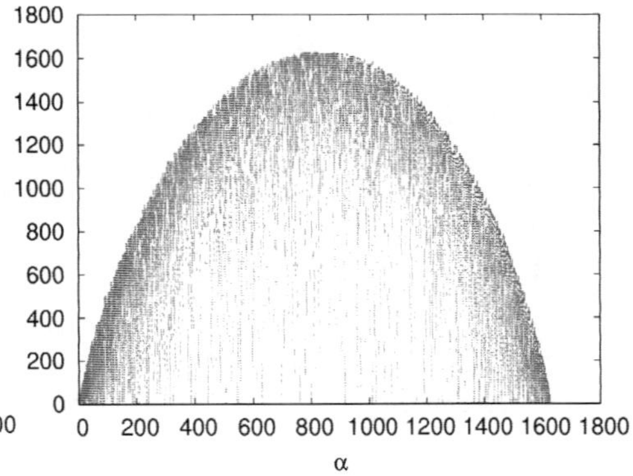

Figure 5. Similar distribution like Fig. 4, for $\chi = 0.18$. The interference pattern is almost erased.

In conclusion, we have studied the statistics of energy levels considering the visibility of its distribution as a function of the parameter of quadrupolar interaction. Visibility is an indication of the degree of coherence of nuclear states associated to studied energies. By comparing the visibility to the respective power spectrum, we found a correspondence between the statistics of energy levels and the degree of purity of the nuclear states. The $1/f$ fluctuations characterizing quantum chaos may be interpreted as a residuum of coherence between nuclear states. The lowest states maintain a high degree of purity, whereas decoherence has almost erased the origin of all intermediate energy states. To define the behavior known as quantum chaos, it is therefore not necessary to manipulate classical or semi-classical images from some dynamic system. Quantum decoherence may be used instead.

Acknowledgments
We would like to thank Etienne Caurier, Andrés Zucker and Frederic Nowacki for stimulating discussions. This work has been supported by DGAPA-UNAM and CONACyT, México.

References

[1] Wojciech H Z 2009 Quantum Darwinism *Nature Physics* **5** 181
[2] Cohen-Tannoudji C, Bernard D and Laloë F 1978 *Quantum Mechanics* (Vol 1) (Paris: Wiley-VCH)
[3] Joos E, Zeh H D, Kiefer C, Giulini D, Kupsch J and Stamatescu I-O 2003 *Decoherence and the Appear-ance of a Classical World in Quantum Theory* (Berlin: Springer)
[4] Zurek W H 2003 Decoherence, einselection, and the quantum origins of the classical *Rev. Mod. Phys.* **75** 715
[5] Schlosshauer M 2007 *Decoherence and the Quantum to Classical Transition* (Berlin: Springer)
[6] Caurier E, Zuker A P, Poves A and Martinez G 1994 *Phys. Rev.* C **50** 225
[7] Dufour M, Zuker A 1996 The realistic collective nuclear Hamiltonian *Phys.Rev.* C **54** 1641
[8] Caurier E and Nowacki F 1998 *Code Antoine* (Strasbourg)
[9] Born M and Wolf E 2001 *Principles of Optics* 7th Ed (Cambridge: Cambridge University Press)
[10] Landa E , Morales I O, Fossion R, Stransky P, Veláquez V, López Vieyra J C and Frank A 2011 *Phys. Rev.* E **84** 016224
[11] Velázquez V, Landa E, Vargas C E, Fossion R, Lopez-Vieyra J C, Morales I and Frank A 2013 Self similitude in the power spectra of nuclear energy levels *J. Phys.: Conf. Ser.* **475** 012014
[12] Hernández-González E M 2012 *Characterizing the photon correlation using the energy spectral density* (Master Thesis) (México: Universidad Nacional Autónoma de México)
[13] Gómez J M G, Kar K. Kota V K B, Molina R A, Relaño A and Retamosa J 2011 *Physics Reports* **499** 103
[14] Faleiro E, Gómez J M G, Molina R A, Munoz L, Relanño A and Retamosa J 2004 *Phys. Rev. Lett.* **93** 244101
[15] Gómez J M G, Relaño A, Retamosa J, Faleiro E, Salasnich L, Vranicar M and Robnik M 2005 *Phys. Rev. Lett.* **94** 084101
[16] Molina R A, Retamosa J, Munñoz L, Relanño A and Faleiro E 2007 *Phys. Lett.* B **644** 25
[17] Molina-Fernández R 2001 *Caos cuántico en sistemas hamiltonianos de muchos cuerpos* (PhD Thesis) (Madrid: Univesidad Complutense de Madrid)
[18] Wigner E P 1967 Random Matrices in Physics *SIAM Rev.* **9(1)** 1
[19] Dyson R J 1962 Statistical Theory of the energy levels of complex systems I, II and III *J. Math. Phys.* **3** 140
[20] Brody T A, Flores J, French J B, Mello P A, Pandey A and Wong S S M 1981 Random-matrix physics: Spectrum and strength fluctuations *Rev. Mod. Phys.* **53** 385
[21] Zelevinsky V, Horoi M and Brown B A 1995 Information entropy, chaos and complexity of the shell model eigenvectors *Phys. Lett.* B **350** 141
[22] Patrick J F, Zelevinsky V, Brown B A, Frazier N and Horoi M 1996 The nuclear shell model as a testing ground for many-body quantum chaos *Phys. Rep.* **276** 85

Journal of Physics: Conference Series **578** (2015) 012015 doi:10.1088/1742-6596/578/1/012015

Valence and sea quarks in the nucleon

Roelof Bijker[1] **and Elena Santopinto**[2]

[1] Instituto de Ciencias Nucleares, Universidad Nacional Autónoma de México, A.P. 70-543, 04510 México, D.F., México

[2] I.N.F.N., Sezione di Genova, via Dodecaneso 33, Genova, I-16146 Italy

E-mail: [1]`bijker@nucleares.unam.mx`

E-mail: [2]`elena.santopinto@ge.infn.it`

Abstract. In this contribution, we discuss the spin and flavor content of the proton in the framework of the unquenched quark model, and address the role of valence and sea quarks in the nucleon.

1. Introduction

The role of valence and sea quarks in the nucleon is addressed in the framework of the unquenched quark model. The constituent quark model (CQM) describes the nucleon as a system of three constituent, or valence, quarks. Despite the successes of the CQM (*e.g.* masses, electromagnetic couplings, magnetic moments), there is compelling evidence for the presence of sea quarks from the measurement of the flavor asymmetry of the proton and the so-called proton spin crisis. The role of the pion cloud in the nucleon has been the subject of many studies [1, 2, 3], and was shown to hold the key to understand the flavor asymmetry and the spin-crisis of the proton. Recently, it was pointed out these two properties are closely related: angular momentum conservation of the pionic fluctuations of the nucleon leads to a relation between the flavor asymmetry and the contribution of orbital angular momentum to the spin of the proton $\mathcal{A}(p) = \Delta L$ [4]. This identity can be understood from the fact that the flavor asymmetry is a matrix element in isospin space, and the orbital angular momentum in spin space with the same values of the quantum numbers.

The aim of this contribution is to study the properties of the nucleon in the unquenched quark model (UQM) at the level of a toy model in which only the effects of the pion cloud is taken into account. It is shown that the pion cloud offers a qualitative understanding of the results obtained in previous numerical studies [5], and thus provides important insights into the properties of the nucleon.

2. Flavor and spin content

In the unquenched quark model the effect of the quark-antiquark pairs is taken into account via a $^3P_0$ creation mechanism. The resulting baryon wave function is given by [5]

$$|\psi_A\rangle = \mathcal{N}\left[|A\rangle + \sum_{BClJ}\int d\vec{K}k^2dk\,\left|BC,l,J;\vec{K},k\right\rangle \frac{\left\langle BC,l,J;\vec{K},k\left|T^\dagger\right|A\right\rangle}{\Delta E_{BC}(k)}\right] , \quad (1)$$

Table 1. Spin and flavor content of the proton in he constituyent quark model (CQM) and the unquenched quark model (UQM).

	CQM	UQM
$\mathcal{A}(p) = \Delta L$	0	$\frac{2a^2-b^2}{3(1+a^2+b^2)}$
Δu	$\frac{4}{3}$	$\frac{4}{3} - \frac{38a^2+b^2-16ab\sqrt{2}}{27(1+a^2+b^2)}$
Δd	$-\frac{1}{3}$	$-\frac{1}{3} + \frac{2a^2+19b^2-16ab\sqrt{2}}{27(1+a^2+b^2)}$
Δs	0	0
$\Delta\Sigma = \Delta u + \Delta d + \Delta s$	1	$1 - \frac{4a^2-2b^2}{3(1+a^2+b^2)}$
$g_A = \Delta u - \Delta d$	$\frac{5}{3}$	$\frac{5}{3} - \frac{40a^2+20b^2-32ab\sqrt{2}}{27(1+a^2+b^2)}$

where $\Delta E_{BC}(k) = M_A - E_B(k) - E_C(k)$ is the energy difference calculated in the rest frame of the initial baryon A with $E_B(k) = \sqrt{M_B^2 + k^2}$ and $E_C(k) = \sqrt{M_C^2 + k^2}$. The operator $T^\dagger$ is the $^3P_0$ quark-antiquark pair creation operator [5, 6]; $\vec{k}$ and l denote the relative radial momentum and orbital angular momentum of B and C, and J is the total angular momentum $\vec{J} = \vec{J}_B + \vec{J}_C + \vec{l}$. The strength of the $^3P_0$ coupling is determined from the flavor asymmetry of the proton.

In this contribution, we employ a simplified version of the UQM in which only the contribution of the pion cloud is taken into account. Table 1 shows the results for the flavor and spin content of the proton. In the UQM, the three coefficients a^2, b^2 and ab are expressed in terms of an integral over the relative momentum k which depends on the $^3P_0$ coupling strength. We note, that the results for the UQM in Table 1 also hold for the meson-cloud model in which the coefficients a and b multiply the $N\pi$ and $\Delta\pi$ components of the nucleon wave function. The ab term denotes the contribution from the cross terms between the $N\pi$ and $\Delta\pi$ components. In the UQM the value of the cross term ab is not equal to the product of a and b, although it turns out that the numerical values are close.

Since the UQM contains the full spin and isospin structure, it satisfies the relation between the flavor asymmetry and the contribution of the orbital angular momentum to the spin of the proton $\mathcal{A}(p) = \Delta L$ [4], and therefore $\Delta\Sigma = 1 - 2\Delta L$. This relation does not hold for the chiral quark model of [7, 8] in which the orbital angular momentum is enhanced with respect to the flavor asymmetry $\Delta L = 3\mathcal{A}(p)/2$ as a consequence of the requirement of a helicity flip of the quark.

Table 2 shows the results for the spin and flavor content of the proton normalized to the proton flavor asymmetry. The third column is normalized to the E866/NuSea value [9], and the fourth column to the somewhat higher NMC value [10]. The experimental values of the spin content were obtained by the HERMES [11] and the COMPASS [13] Collaborations. In Table 2, we show the HERMES results.

The probability that a proton fluctuates in $n\pi^+$

$$\left|\langle n\pi^+|p\rangle\right|^2 = \frac{2a^2}{3(1+a^2+b^2)} = 0.180 \; , \tag{2}$$

(UQM1 value) is in close agreement with the experimental value 0.17 ± 0.01 determined in an analysis of forward neutron production in electron-proton collisions at 300 GeV by the H1 and ZEUS Collaborations at DESY [14, 15]. The UQM2 value is somewhat higher 0.241. The total

Table 2. Spin and flavor content of the proton normalized to the flavor asymmetry, UQM1 using the E866/NuSea value [9] and UQM2 using the NMC value [10].

	CQM	UQM1	UQM2	Exp	Ref
$\mathcal{A}(p)$	0	$*0.118$	$*0.158$	0.118 ± 0.012	[9]
				0.158 ± 0.010	[10]
Δu	4/3	1.132	1.064	0.842 ± 0.013	[11]
Δd	$-1/3$	-0.368	-0.380	-0.427 ± 0.013	[11]
Δs	0	0	0	-0.085 ± 0.018	[11]
$\Delta\Sigma$	1	0.764	0.684	0.330 ± 0.039	[11]
g_A	5/3	1.500	1.444	1.2701 ± 0.0025	[12]

probability for a pion fluctuation of the proton is given by

$$|\langle N\pi|p\rangle|^2 + |\langle \Delta\pi|p\rangle|^2 = \frac{a^2+b^2}{1+a^2+b^2} = 0.455 , \tag{3}$$

(UQM1 value), in good agreement with the value of 0.470 as determined in an analysis of the quark distribution functions measured in Drell-Yan experiments and semi-inclusive DIS experiments [16]. Also in this case, the UQM2 value, 0.609, is about 30 % higher than the UQM1 value.

3. Summary and conclusions

In this contribution, we studied the properties of the proton in the framework of the unquenched quark model in which the $^3P_0$ coupling strength was normalized to the observed value of the proton flavor asymmetry. It was shown that the pion fluctuations help to understand the discrepancies between the constituent quark model and the experimental data. Their inclusion leads to a reduction of quark model value of Δu and g_A, and give rise to a sizeable contribution (25 - 30 %) of orbital angular momentum to the spin of the proton. In addition, it was found that the probabilities for pion fluctuations in the UQM are in good agreement with the values determined in analyses of the available experimental data.

Acknowledgments

This work was supported in part by research grants from CONACyT and PAPIIT-UNAM.

References

[1] Kumano S 1998 *Phys. Rep.* **303** 183
[2] Speth J and Thomas A W 1998 *Adv. Nucl. Phys.* **24** 83
[3] Garvey G T and Peng J C 2001 *Prog. Part. Nucl. Phys.* **47** 203
[4] Garvey G T 2010 *Phys. Rev.* C **81** 055212
[5] Bijker R and Santopinto E 2009 *Phys. Rev.* C **80** 065210
Santopinto E and Bijker R 2010 *Phys. Rev.* C **82** 062202(R)
Bijker R, Ferretti J and Santopinto E 2012 *Phys. Rev.* C **85** 035204
[6] Roberts W and Silvestre-Brac B 1992 *Few-Body Systems* **11** 171
[7] Eichten E J, Hinchcliffe I and Quigg C 1992 *Phys. Rev.* D **45** 2269
[8] Cheng T P and Li L F 1995 *Phys. Rev. Lett.* **74** 2872
[9] Towell R S *et al.* (FNAL E866/NuSea Collaboration) 2001 *Phys. Rev.* D **64** 052002
[10] Arneodo M *et al.* (New Muon Collaboration) 1997 *Nucl. Phys.* B **487** 3

[11] Airapetian A *et al.* (HERMES Collaboration) 2007 *Phys. Rev.* D **75** 012007
[12] Beringer J *et al.* (Particle Data Group) 2012 *Phys. Rev.* D **86**, 010001
[13] Alexakhin V Yu *et al.* (COMPASS Collaboration) 2007 *Phys. Lett.* B **647** 8
[14] Bunyatyan A and Povh B 2006 *Eur. Phys. J.* A **27** 359
[15] Povh B and Rosina M 2011 *Bled Workshops in Physics* **12** 82
[16] Chang W C and Peng J C 2011 *Phys. Rev. Lett.* **106** 252002

Journal of Physics: Conference Series **578** (2015) 012016 doi:10.1088/1742-6596/578/1/012016

Coulomb gauge approach for charmonium meson and hybrid radiative transitions.

Peng Gou[1,2], Tochtli Yépez-Martínez[2], Adam P Szczepaniak[1,2,3]

[1] Thomas Jefferson National Accelerator Facility, Newport News, VA 23606, USA

[2] Center For Exploration of Energy and Matter, Indiana University, Bloomington, IN 47408, USA.

[3] Physics Department, Indiana University, Bloomington, IN 47405, USA

E-mail: pguo@jlab.org, tyepezma@indiana.edu, aszczepa@indiana.edu

Abstract.
We consider the lowest order interaction of the Foldy-Wouthuysen QED and QCD Hamiltonian in the Coulomb gauge approach, to describe radiative transitions between conventional and hybrids charmonium mesons. The results are compared to potential quark models and lattices calculations.

1. Introduction

It has long been expected that excitation of the gluon field should contribute to the spectrum of hadrons. Hybrid resonances, *i.e.* states that contain both quark and gluon excitations, were considered in various models, [1, 2, 3, 4, 5, 6], and recent lattice simulations [7, 8, 9] confirm model expectations. Furthermore, in recent years several new resonances, in particular in the charmonium spectrum have been identified and may include hybrids, *e.g.* the $Y(4260)$. Conventional heavy quarkonia are well described by non-relativistic QCD [10]. Thus it is reasonable to expect that hybrids containing heavy quarks could be treated in a similar way, *i.e.* by considering gluon excitations in presence of slowly moving quarks.

To a good approximation heavy quarks interact with photons as bare Dirac particles. Thus radiative transitions can be used to explore quarkonium dynamics. We assume that this phenomenology may be extended to quarkonium hybrids [11, 12]. Over the years several radiative transitions have been measured [13, 14, 15] and extensive theoretical studies were performed [16, 17, 18, 19]. More recently lattice gauge simulations have become available [20, 21] that also include predictions for transitions involving hybrid mesons [22, 23].

In this contribution we present the radiative transitions involving the lowest mass conventional charmonia and the lowest mass multiplet of charmonium hybrids. We also discuss current matrix elements involving states of identical charge conjugation. These vanish when photon couples to both the quark and the antiquark but are in general finite when the current operator acts on a single quark. They are well defined within the model and can also be computed on the lattice.

2. Coulomb gauge approach and the basic ingredients

The QCD Hamiltonian H_{QCD}, which describes non-relativistic quarks interacting with (relativistic) gluons can be constructed from the full QCD Hamiltonian in the Coulomb gauge by applying Foldy-Wouthuysen transformation [24]. This Hamiltonian was used to study the gluelump spectrum [11] and the low mass charmonia and bottomonia including hybrids [12]. In addition to the strong interaction part, here we also consider the minimal coupling of the photon to the quarks, which in the non-relativistic limit is given by

$$H_{QED} = \frac{e_q}{2m}\int d\mathbf{x}\Psi^\dagger(\mathbf{x})\beta[2i\mathbf{A}_\gamma(\mathbf{x})\cdot\nabla - \mathbf{\Sigma}\cdot\mathbf{B}_\gamma(\mathbf{x})]\Psi(\mathbf{x}) \tag{1}$$

where $\mathbf{A}_\gamma$ and $\mathbf{B}_\gamma$ are the photon vector potential and magnetic field, respectively. The quark fields are given by

$$\Psi^i(\mathbf{x}) = \sum_{\lambda=\pm 1/2}\int \frac{d\mathbf{k}}{(2\pi)^3}e^{i\mathbf{k}\cdot\mathbf{x}}[u_\lambda b(\mathbf{k},\lambda,i) + v_\lambda d^\dagger(-\mathbf{k},\lambda,i)] \tag{2}$$

with u, v being the Dirac spinors in the non-relativistic limit. Given an approximate solution of the Schrödingier equation $H_{QCD}|N[c\bar{c}]\rangle = E_N|N[c\bar{c}]\rangle$, within the Fock sector containing only the heavy quark-antiquark pair, the QED interaction of Eq. (1) determines the radiative transition matrix element,

$$\mathcal{M}_{N\to N'\gamma} \propto \langle N'[c\bar{c}],\gamma|H_{QED}|N[c\bar{c}]\rangle \tag{3}$$

between ordinary charmonia. In the case of transitions involving hybrids, which are given by solutions of $H_{QCD}|N[c\bar{c}g]\rangle = E_N|N[c\bar{c}g]\rangle$ in the sector containing in addition to the $c\bar{c}$ pair a transverse quasi-gluon, the radiative transition to an ordinary meson state has to be accompanied by gluon absorption. To lowest order in the heavy quark mass expansion the later is determined by the instantaneous Coulomb interaction that changes the gluon number, $\langle c\bar{c}|H_C|c\bar{c}g\rangle$. Here H_C is given by

$$H_C = -\frac{g^2}{2}\int d\mathbf{x}d\mathbf{y}\rho^a(\mathbf{x})K_{a,b}(\mathbf{x},\mathbf{y},\mathbf{A}_g)\rho^b(\mathbf{y}) \tag{4}$$

and $\rho^a(\mathbf{x}) = \Psi^\dagger(\mathbf{x})T^a\Psi(\mathbf{x})$ is the quark color charge density and the gluon field $\mathbf{A}_g$ is related to the quasi-gluon particle operators by

$$\mathbf{A}_g^a(\mathbf{x}) = \int\frac{d\mathbf{k}}{(2\pi)^3}\frac{e^{i\mathbf{k}\cdot\mathbf{x}}}{\sqrt{2\omega(k)}}[\vec{\epsilon}(\mathbf{k},\lambda)a(\mathbf{k},\lambda,a) + \vec{\epsilon}^\dagger(-\mathbf{k},\lambda)a^\dagger(-\mathbf{k},\lambda,a)], \tag{5}$$

with $\omega(k)$ the quasi-gluons dispersion function, (λ, a) being the helicity and color indices, respectively and $\vec{\epsilon}(\mathbf{k},\lambda)$ the helicity vectors. In the variational model the Coulomb kernel is replaced by its vacuum expectation value and the operator which changes the gluon number by one becomes,

$$K_{a,b} = f^{abc}\int\frac{d\mathbf{k}}{(2\pi)^3}\frac{d\mathbf{q}}{(2\pi)^3}e^{i\mathbf{kx}-i\mathbf{q}\cdot\mathbf{y}}\mathbf{k}\cdot\mathbf{A}_g^c(\mathbf{k}-\mathbf{q})K^1(k,q) \tag{6}$$

with the scalar function $K^1(k,q)$ obtained from a solution of a series of Dyson-Schwinger equations [25, 26, 27, 28, 29, 30, 31]. The model has been used successfully [32, 33] in the study of excited adiabatic potentials between static quarks [34], which can be used to determine the single gluon orbitals in Eq. (5). Combining Eqs. (1,4) leads to an effective operators for radiative transitions between hybrid and ordinary quarkonia

$$\mathcal{M}_{N\to N'\gamma} \propto \langle N'[c\bar{c}],\gamma|H_{QED}^{eff}|N[c\bar{c}g]\rangle \tag{7}$$

where $H_{QED}^{eff} = \frac{H_C H_{QED}}{\Delta E}$, with $1/\Delta E$ representing the Green's function of the $c\bar{c}$ pair. In the following we calculate the decay widths and transition amplitudes for several mesons and hybrid states.

2.1. Meson basis.

We represent the N-th quarkonium state of spin and its projection J, M, with parity P and charge conjugation C and total momentum $\mathbf{P}$ by

$$|\mathbf{P}; JMPCN\rangle = \sum_{\alpha,m_1,m_2} \int \frac{d\mathbf{q}}{(2\pi)^3} \Psi^{N,\alpha}_{c\bar{c}}(q) \chi^{JMPC}_{m_1,m_2}(\hat{\mathbf{P}}, \hat{\mathbf{q}}, \alpha) b^\dagger(\mathbf{p}_c, m_1, i_1) \frac{\delta_{i_1,i_2}}{\sqrt{N_c}} d^\dagger(\mathbf{p}_{\bar{c}}, m_2, i_2)|0\rangle. \quad (8)$$

Here $\alpha = (L, S)$, and q is the magnitude of relative momentum between quark and antiquark. $\mathbf{p}_c = \frac{\mathbf{P}}{2} + \mathbf{q}$ and $\mathbf{p}_{\bar{c}} = \frac{\mathbf{P}}{2} - \mathbf{q}$ are the quark and antiquark momenta respectively. The meson spin-orbital wave function is written using the $L - S$ coupling scheme with L, and S the orbital angular momentum and spin of the quark-antiquark, respectively,

$$\begin{aligned} \chi^{JMPC}_{m_1,m_2}(\hat{\mathbf{P}}, \hat{\mathbf{q}}, \alpha) &= \sum_{M_S,M_L} Y_{LM_L}(\hat{\mathbf{q}}) \langle \frac{1}{2} m_1; \frac{1}{2} m_2 | SM_S \rangle \langle SM_S; LM_L | JM \rangle \\ &\times \frac{1 + C(-1)^{L+S}}{2} \frac{1 + P(-1)^{L+1}}{2}. \end{aligned} \quad (9)$$

We consider the spin-orbital wave function $\chi^{JMPC}_{m_1,m_2}(\hat{\mathbf{q}}, \alpha)$ for the charmonium mesons $\eta_c(0^{-+})$, $J/\psi(1^{--})$, $h_c(1^{+-})$, $\chi_{c0}(0^{++})$, $\chi_{c1}(1^{++})$ and $\chi_{c2}(2^{++})$. The states are normalized according to $\langle \mathbf{P}'; J'M'P'C'N' | \mathbf{P}; JMPCN \rangle = 2E_{c\bar{c}}(2\pi)^3 \delta^3(\mathbf{P} - \mathbf{P}') \delta_{JJ'} \delta_{MM'} \delta_{PP'} \delta_{CC'} \delta_{NN'}$.

2.2. Hybrid basis

In construction of hybrid wave functions we thus follow the coupling scheme optimized for gluelump studies [11]. The hybrid state with total spin and it projection, J,M, parity P, charge conjugation C is then given by

$$\begin{aligned} |JMPCN\rangle &= \sum_{\alpha=(J_g S,L,j)} \int \frac{d\mathbf{k}}{(2\pi)^3} \frac{d\mathbf{q}}{(2\pi)^3} \Psi^{N,\alpha}_{c\bar{c}g}(k,q) \sum_{m_1,m_2,\sigma} \frac{1}{\sqrt{C_F N_c}} \chi^{JMPC}_{m_1,m_2,\sigma}(\hat{\mathbf{k}}, \hat{\mathbf{q}}, \alpha) \\ &\times b^\dagger(\frac{\mathbf{k}}{2} + \mathbf{q}, m_1, i_1) T^a_{i_1,i_2} d^\dagger(\frac{\mathbf{k}}{2} - \mathbf{q}, m_2, i_2) a^\dagger(-\mathbf{k}, \sigma, a)|0\rangle. \end{aligned} \quad (10)$$

The spin-orbital wave function $\chi^{JMPC}_{m_1,m_2,\sigma}(\hat{\mathbf{k}}, \hat{\mathbf{q}}, \alpha)$ describes the $(\boldsymbol{L} + \boldsymbol{J}_g) + \boldsymbol{S}$ coupling, with J_g the total spin of the gluon and $\sigma = \pm 1$ the gluon helicity

$$\begin{aligned} \chi^{JMPC}_{m_1,m_2,\sigma}(\hat{\mathbf{k}}, \hat{\mathbf{q}}, \alpha) = \sqrt{\frac{2J_g + 1}{4\pi}} \frac{1 + C(-1)^{L+S+1}}{2} \frac{(-1)^{J_g}}{\sqrt{2}} \sum_{M_g,M_S,M_L,m} Y_{LM_L}(\mathbf{q}) \langle \frac{1}{2} m_1, \frac{1}{2} m_2 | SM_S \rangle \\ \times \langle J_g M_g, LM_L | jm \rangle \langle jm, SM_S | JM \rangle D^{*J_g}_{M_g,-\sigma}(\hat{\mathbf{k}}) [\delta_{\sigma,1} + P(-1)^{J_g+L+1} \delta_{\sigma,-1}] \end{aligned} \quad (11)$$

Here $\mathbf{q}$ is the relative momentum between the quark-antiquark and $\mathbf{k}$ is the momentum of the gluon in the overall center of mass frame. The parity and charge conjugation are given by $P = \xi(-1)^{J_g+L+1}$,$C = (-1)^{L+S+1}$, respectively. Here $\xi = +1$ corresponds to the TM (natural parity) and $\xi = -1$ for TE (unnatural parity) gluon state that are given by $|\sigma = +1\rangle + \xi|\sigma = -1\rangle$ combinations of gluon helicity states. The state is normalized in the same way as the normalization of conventional meson state.

For the lowest four hybrids we are considering [12], $(L, J_g^{P_g C_g}) = (0, 1^{+-})$, which correspond to the gluon in the TE mode. Coupling the TE gluon with the color octet $Q\bar{Q}$ state in $L = 0$, produces a hybrid state with the intermediate angular momentum $\mathbf{j} = \mathbf{L} + \mathbf{J}_g = 1$. Adding the quark spin $S = 0, 1$, and ignoring hyperfine splitting we obtain four low lying hybrids with quantum numbers, $J^{PC} = Y(1^{--})$ for $S = 0$ and $J^{PC} = 0^{-+}, \eta_{c1}(1^{-+}), 2^{-+}$ for $S = 1$. It is worth noting that the hybrid with exotic quantum numbers 1^{-+} appears in this lowest multiplet and is predicted to have the $Q\bar{Q}$ pair in spin-1. These states has been described by lattice calculations [20, 22] and is expected to have a mass around 4.3 GeV.

3. Decay widths, transition amplitudes and predictions.

3.1. Meson-to-meson radiative transitions

We have considered fifteen transitions between conventional charmonia. Even though some of the transitions considered here vanish due to charge conjugation, we investigate the underlying matrix elements with photon attached to only one of the quarks. Some of these C-violating results can be compared with lattice results reported in [22], and others constitute our predictions. The matrix elements are obtained by using the model described in Sec. 2 and decay widths computed from

$$\Gamma(N \to N'\gamma) = \int d\Omega_\gamma \frac{1}{32\pi^2}\frac{k_\gamma}{m_N^2}\frac{1}{2J_N+1}\sum_{\sigma_\gamma, M_N, M_{N'}} |\mathcal{M}_{N\to N'\gamma}|^2. \tag{12}$$

The photon momentum for each transition is given by $k_\gamma = (M_N^2 - M_{N'}^2)/2M_N$. For the radial wave functions we use a harmonic oscillator approximation with a width parameter $\beta = 0.5$ GeV. This leads to some differences with respect to the potential-quark results of [19], where a Coulomb plus linear plus and hyperfine interactions were used to compute the wave functions. In Figure 1, our Coulomb gauge approach is compared with PDG reports [35] and non-relativistic (NR) potential-quark results [19].

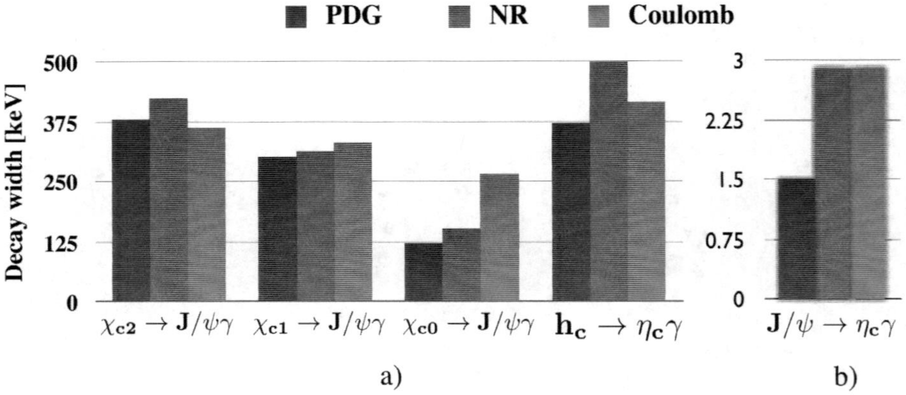

Figure 1. Decay widths for $c\bar{c}$-mesons, a) electric and b) magnetic dipole transitions.

We calculate the transition amplitudes $TA = |\hat{V}|$, $|\hat{F}_1|$ introduced in [20, 22] in the context of analysis of lattice data. Here $\hat{F}_1$ represents either electric, $\hat{E}_1$ or magnetic, $\hat{M}_1$ dipole transition and $\hat{V}$ is the magnetic dipole transition involving a vector and a pseudoscalar meson,

$$|\hat{F}_1|^2 = \frac{1}{8e_q^2}\sum_{\sigma_\gamma, M_N, M_{N'}} |\mathcal{M}_{N\to N'\gamma}|^2, \qquad |\hat{V}|^2 = \frac{(m_N + m_{N'})^2}{32e_q^2 m_N^2 k_\gamma^2}\sum_{\sigma_\gamma, M_N, M_{N'}} |\mathcal{M}_{N\to N'\gamma}|^2. \tag{13}$$

In Figure 2, we compare the transition amplitudes in the Coulomb gauge approach with the available lattice computations [20, 22].

Furthermore, we found other three non-zero transitions, not reported so far, which decay widths and TA correspond to , $\Gamma(\chi_{c2} \to h_c\gamma) = 0.1$ keV, $|\hat{F}| = 0.12$ GeV, $\Gamma(h_c \to \chi_{c1}\gamma) = 0.001$ keV, $|\hat{F}| = 0.01$ GeV and $\Gamma(h_c \to \chi_{c0}\gamma) = 0.6$ keV, $|\hat{F}| = 0.13$ GeV, these constitute our predictions. Finally, we calculate the transition amplitudes for charge conjugation violating transitions, $|\hat{F}|/2$. The factor of two is introduced to account for the fact that photon couples to

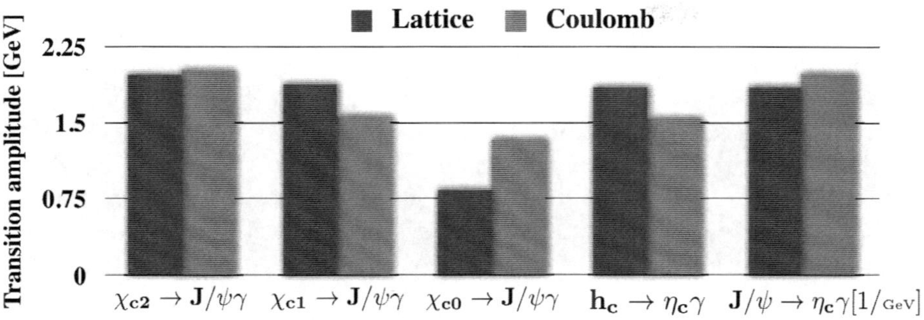

Figure 2. Transition amplitudes for $c\bar{c}$-mesons.

a single quark. We found two C-violating transitions $\chi_{c2} \to \chi_{c1}\gamma$ and $\chi_{c1} \to \chi_{c0}\gamma$ and their TA are given by $\hat{F} = 0.10$ GeV and $\hat{F} = 0.17$ GeV, respectively, which also constitute predictions of the model.

3.2. Hybrid-to-meson radiative transitions

We have studied 24 possible hybrid to meson radiative transitions, including matrix elements for C-violating modes, using $m_{hyb} = 4.35$ GeV for the spin-averaged mass of the lowest hybrid multiplet $Y(1^{--}), 0^{-+}, \eta_{c1}(1^{-+}), 2^{-+}$.

Lattice simulations [22] reports two decay widths $\Gamma(Y \to \eta_c\gamma) = 42$ keV, $\Gamma(\eta_{c1} \to J/\psi\gamma) = 115$ keV and both correspond to a magnetic dipole transition. These decay widths are obtained from

$$\Gamma(Y \to \eta_c\gamma) = \alpha k_\gamma^3 \frac{64}{27}\frac{|\hat{V}|^2}{(m_Y + m_{\eta_c})^2} \quad \text{and} \quad \Gamma(\eta_{c_1} \to J/\psi\gamma) = \alpha k_\gamma \frac{16}{27}\frac{|\hat{F}|^2}{m_{\eta_{c1}}^2} \tag{14}$$

with $\hat{V} = 0.28$ and $\hat{F} = 0.69$ GeV, respectively.

We found four non-zero hybrid to meson transition amplitudes in our model, *i.e.*, $Y \to \eta_c\gamma$, $0^{-+} \to J/\psi\gamma, \eta_{c1} \to J/\psi\gamma$ and $2^{-+} \to J/\psi\gamma$. All of them depend on a single factor that is determined by the hybrid meson wave function. Therefore, the four decay widths can be written as $\Gamma = \alpha k_\gamma^3 \kappa_1^2$ or $\Gamma = \alpha k_\gamma \kappa_2^2$, showing explicitly a symmetry relation implied by the light hybrid multiplet. The common factors κ_1 and κ_2 can be normalized by the two magnetic dipole matrix elements $\hat{V}$ and $\hat{F}$, respectively. To minimize sensitivity to the hybrid wave functions we calculate the ratio of hybrid decay widths. In general the model predicts $\mathcal{R} = \frac{\Gamma(N \to N'\gamma)}{\Gamma(\eta_{c1}(1^{-+}) \to J/\psi\gamma)} \approx 1$ for the whole hybrid multiplet while lattice reports $\mathcal{R} = \frac{\Gamma(Y(1^{--}) \to \eta_c\gamma)}{\Gamma(\eta_{c1}(1^{-+}) \to J/\psi\gamma)} = 0.37$.

For the C-violating hybrid to meson matrix elements $\mathcal{M}_{CV}$ we found that all of them are proportional to a common factor *i.e.*, $\mathcal{M}_{CV} \propto \kappa_3$. We use the transition amplitude $|\hat{E}|_{1^{-+} \to 0^{++}\gamma} = 0.34$ GeV from [22] to normalize the relevant wave function overlap factor κ_3 in our model. Therefore, we predict seven non-zero C-violating transitions amplitudes $|\hat{E}_1|$ within the range of $0.30 - 0.66$ GeV and correspond to the transitions $Y \to h_c\gamma$, $0^{-+} \to \chi_{c1}\gamma$, $\eta_{c1} \to \chi_{c0}\gamma$, $\eta_{c1} \to \chi_{c1}\gamma$, $\eta_{c1} \to \chi_{c2}\gamma$, $2^{-+} \to \chi_{c1}\gamma$ and $2^{-+} \to \chi_{c2}\gamma$.

4. Summary

We studied radiative decays of conventional charmonia and charmonium hybrids. We used the ordinary $c\bar{c}$-mesons $\eta_c(0^{-+}), J/\psi(1^{--}), \chi_{c0}(0^{++}), \chi_{c1}(1^{++}), h_c(1^{+-}), \chi_{c2}(2^{++})$, where we considered the minimal coupling of the photon to the non-relativistic quarks. We have compared our results with non-relativistic quark model and lattice calculations and found a reasonable agreement. A few new predictions for transition amplitudes were presented including charge violating transitions amplitudes.

To describe hybrid decays we considered a model based on an effective QCD Hamiltonian that describes non-relativistic quarks interacting with (relativistic) gluons and is constructed from the QCD in the Coulomb gauge by applying Foldy-Wouthuysen transformation. We considered decays of states from the hybrid multiplet $1^{--}, (0;1;2)^{-+}$ with 1^{-+} being the exotic state. There are 24 possible radiative transitions between this multiplet and ground state charmonia. We found four non-zero decay widths for which the model predicts $\mathcal{R} = \frac{\Gamma(N \to N'\gamma)}{\Gamma(\eta_{c1}(1^{-+}) \to J/\psi\gamma)} \approx 1$ for the whole hybrid multiplet while lattice reports $\mathcal{R} = \frac{\Gamma(Y(1^{--}) \to \eta_c\gamma)}{\Gamma(\eta_{c1}(1^{-+}) \to J/\psi\gamma)} = 0.37$. We also investigated C-violating matrix elements involving hybrids. The model predicts several of such matrix elements to be non-zero and of the same order of magnitude as the reported transition $1^{-+} \to 0^{++}\gamma$ by lattice calculations.

Acknowledgements
We acknowledge financial help from National Research Council of Mexico (CONACyT), the US. Department of Energy, and Indiana University Collaborative Research Grant.

References

[1] Horn D and Mandula J 1978 *Phys. Rev.* D **17** 898
[2] Isgur N and Paton J E 1985 *Phys. Rev.* D **31** 2910
[3] Simonov Y A 2001 *Nucl. Phys.* B **592** 350
[4] Szczepaniak A, Swanson E S, Ji C R and Cotanch S R 1996 *Phys. Rev. Lett.* **76** 2011
[5] Buisseret F and Semay C 2006 *Phys. Rev.* D **74** 114018
[6] Brau F and Semay C 2004 *Phys. Rev.* D **70** 014017
[7] Morningstar C J and Peardon M J 1999 *Phys. Rev.* D **60** 034509
[8] Foster M and Michael C, 1999 *Phys. Rev.* D **59** 094509
[9] Bali G S and Pineda A 2004 *Phys. Rev.* D **69** 094001
[10] Eichten E and Feinberg F 1981 *Phys. Rev.* D **23** 2724
[11] Guo P, Szczepaniak A P, Galata G, Vassallo A and Santopinto E 2008 *Phys. Rev.* D **77** 056005
[12] Guo P, Szczepaniak A P, Galata G, Vassallo A and Santopinto E 2008 *Phys. Rev.* D **78** 056003
[13] Crede V and Meyer C A 2009 *Prog. Part. Nucl. Phys.* **63** 74
[14] Mathieu V Kochelev N and Vento V 2009 *Int. J. Mod. Phys.* E **18** 1
[15] Swanson E S 2006 *Phys. Rep.* **429** 243
[16] Eichten E, Gottfried K, Kinoshita T, Lane K D and Yan T M 1978 *Phys. Rev.* D **17** 3090
[17] Eichten E, Gottfried K, Kinoshita T, Lane K D and Yan T M 1980 *Phys. Rev.* D **21** 203
[18] Eichten E J, Lane K and Quigg C 2002 *Phys. Rev. Lett.* **89** 162002
[19] Barnes T, Godfrey S and Swanson E S 2005 *Phys. Rev.* D **72** 054026
[20] Dudek J J, Edwards R G and Richards D G 2006 *Phys. Rev.* D **73** 074507
[21] Dudek J J, Edwards R G, Mathur N and Richards D G 2008 *Phys. Rev.* D **77** 034501
[22] Dudek J J, Edwards R G and Thomas C E 2009 *Phys. Rev.* D **79** 094504
[23] Dudek J J 2001 *Phys. Rev.* D **84** 074023
[24] Feinberg F L 1978 *Phys. Rev.* D **17** 2659
[25] Szczepaniak A P and Swanson E S 2001 *Phys. Rev.* D **65** 025012
[26] Szczepaniak A P 2004 *Phys. Rev.* D **69** 074031
[27] Feuchter C and Reinhardt H 2004 *Phys. Rev.* D **70** 105021
[28] Reinhardt H and Feuchter C 2005 *Phys. Rev.* D **71** 105002
[29] Schleifenbaum W, Leder M and Reinhardt H 2006 *Phys. Rev.* D **73** 125019
[30] Epple D, Reinhardt H and Schleifenbaum W 2007 *Phys. Rev.* D **75** 045011
[31] Epple D, Reinhardt H, Schleifenbaum W and Szczepaniak A P 2008 *Phys. Rev.* D **77** 085007
[32] Szczepaniak A P and Krupinski P 2006 *Phys. Rev.* D **73** 116002
[33] Szczepaniak A P and Krupinski P 2006 *Phys. Rev.* D **73** 034022
[34] Juge K J, Kuti J and Morningstar C J 1998 *Nucl. Phys. B: Proc. Suppl.* **63**, 326
[35] J. Beringer et al. (Particle Data Group) 2012 *Phys. Rev.* D **86** 010001

Institute of Physics
Dirac House, Temple Back
Bristol BS1 6BE UK

CURRAN ASSOCIATES INC
proceedings
.com